Barbara Leitenmaier

Metal transport in the hyperaccumulator plant Thlaspi caerulescens

Barbara Leitenmaier

Metal transport in the hyperaccumulator plant Thlaspi caerulescens

Transport and detoxification of cadmium, copper and zinc in the Cd/Zn hyperaccumulator plant Thlaspi caerulescens

Südwestdeutscher Verlag für Hochschulschriften

Impressum/Imprint (nur für Deutschland/only for Germany)
Bibliografische Information der Deutschen Nationalbibliothek: Die Deutsche Nationalbibliothek verzeichnet diese Publikation in der Deutschen Nationalbibliografie; detaillierte bibliografische Daten sind im Internet über http://dnb.d-nb.de abrufbar.
Alle in diesem Buch genannten Marken und Produktnamen unterliegen warenzeichen-, marken- oder patentrechtlichem Schutz bzw. sind Warenzeichen oder eingetragene Warenzeichen der jeweiligen Inhaber. Die Wiedergabe von Marken, Produktnamen, Gebrauchsnamen, Handelsnamen, Warenbezeichnungen u.s.w. in diesem Werk berechtigt auch ohne besondere Kennzeichnung nicht zu der Annahme, dass solche Namen im Sinne der Warenzeichen- und Markenschutzgesetzgebung als frei zu betrachten wären und daher von jedermann benutzt werden dürften.

Verlag: Südwestdeutscher Verlag für Hochschulschriften GmbH & Co. KG
Heinrich-Böcking-Str. 6-8, 66121 Saarbrücken, Deutschland
Telefon +49 681 37 20 271-1, Telefax +49 681 37 20 271-0
Email: info@svh-verlag.de

Approved by: Konstanz, Universität, Diss., 2010

Herstellung in Deutschland:
Schaltungsdienst Lange o.H.G., Berlin
Books on Demand GmbH, Norderstedt
Reha GmbH, Saarbrücken
Amazon Distribution GmbH, Leipzig
ISBN: 978-3-8381-3050-7

Imprint (only for USA, GB)
Bibliographic information published by the Deutsche Nationalbibliothek: The Deutsche Nationalbibliothek lists this publication in the Deutsche Nationalbibliografie; detailed bibliographic data are available in the Internet at http://dnb.d-nb.de.
Any brand names and product names mentioned in this book are subject to trademark, brand or patent protection and are trademarks or registered trademarks of their respective holders. The use of brand names, product names, common names, trade names, product descriptions etc. even without a particular marking in this works is in no way to be construed to mean that such names may be regarded as unrestricted in respect of trademark and brand protection legislation and could thus be used by anyone.

Publisher: Südwestdeutscher Verlag für Hochschulschriften GmbH & Co. KG
Heinrich-Böcking-Str. 6-8, 66121 Saarbrücken, Germany
Phone +49 681 37 20 271-1, Fax +49 681 37 20 271-0
Email: info@svh-verlag.de

Printed in the U.S.A.
Printed in the U.K. by (see last page)
ISBN: 978-3-8381-3050-7

Copyright © 2012 by the author and Südwestdeutscher Verlag für Hochschulschriften GmbH & Co. KG and licensors
All rights reserved. Saarbrücken 2012

Table of contents

Table of contents 1

Summary 3

Zusammenfassung 5

1. Introduction 9
   1.1. State of the art 9
   1.2. Aims and objectives of this thesis 11

2. Publications in peer-reviewed journals and manuscripts 15
   2.1. Cadmium-induced inhibition of photosynthesis and long-term acclimation to cadmium stress in the Cd hyperaccumulator *Thlaspi caerulescens* **published in New Phytologist (2007) 175: 655-674** 15
   2.2. Cadmium uptake and sequestration kinetics in individual leaf cell protoplasts of the Cd/Zn hyperaccumulator *Thlaspi caerulescens* **published in Plant, Cell & Environment (2011) 34: 208-219** 51
   2.3. A native Zn/Cd pumping P_{1B} ATPase from natural overexpression in a hyperaccumulator plant **published in BBRC (2007) 363: 51-56** 75
   2.4. Biochemical and biophysical characterisation yields insights into the mechanism of a Cd/Zn transporting ATPase purified from the hyperaccumulator plant *Thlaspi caerulescens* **published in Biochimica et Biophysica Acta (section Biomembranes) 1808: 2591-2599** 85
   2.5. Zn EXAFS and UV/Vis spectroscopy of TcHMA4 – preliminary results 107
   2.6. Complexation and toxicity of copper in higher plants (II): Different mechanisms for Cu vs. Cd detoxification in the Cu-sensitive Cd/Zn hyperaccumulator *Thlaspi caerulescens* (Ganges ecotype) **published in Plant Physiology (2009) 151: 715-731** 111

3. General discussion 145
4. References 151
5. Appendix 167

5.1. Protocol for isolation and purification of native TcHMA4 from *Thlaspi caerulescens* roots 167

5.2. Author contributions 181

5.3. Acknowledgements 183

Summary

In this thesis, various aspects on heavy metal accumulation by the hyperaccumulator plant *Thlaspi caerulescens* have been investigated. *T. caerulescens* belongs to the family of Brassicaceae and hyperaccumulates zinc. Its ecotype Ganges, originating from Southern France, additionally takes up cadmium actively. It is known from previous studies that hyperaccumulators have highly overexpressed metal transporters and that most of them store the metal in the vacuole of large epidermal cells.

Cd acclimation and sequestration in *Thlaspi caerulescens*

First, the long-term behaviour of *T. caerulescens* upon cadmium treatment has been studied. For this purpose, plants were grown for six months on a nutrient solution containing elevated concentrations of cadmium. First, they showed toxicity symptoms like yellowing of leaves, but continued growing. After two months, the plants started to acclimate and toxicity symptoms almost disappeared. Using chlorophyll fluorescence kinetic measurements it has been shown that during acclimation, not all cells are affected by cadmium. The distribution of cadmium within the leaves was heterogenous, some mesophyll cells took up much more metal than others. Slowly this heterogenity disappeared with the metal being sequestred into epidermal vacuoles. The study also showed that cadmium inhibits the photosynthetic light reactions more than the Calvin-Benson cycle and that at least two different targets in/around photosystem II are affected by cadmium. Using a fluorescent dye specific for cadmium and protoplasts from *Thlaspi* leaves, we were able to show cadmium uptake into mesophyll cells as well as normal sized and storage epidermal cells. The uptake rates into storage cells were significantly higher than the uptake rates into mesophyll or normal sized epidermal cells. This shows that the differential accumulation in leaf tissues is not due to differences in cell walls or transpiration stream (absent in protoplasts), but different expression levels of transport proteins. Shortly after addition of cadmium to the measuring medium, a bright ring inside the cells appeared and stayed there for some time. Very slowly the whole cell became bright, showing that the sequestration from the cytoplasm into the vacuole is one time limiting step in cadmium hyperaccumulation in *T. caerulescens*.

Metal transport and detoxification

Not much is known about metal transporters in plants in general and about metal transporting ATPases in particular. As metal ATPases play an important role in hyperaccumulation, TcHMA4, a P_{1B}-type ATPase that is suggested to pump cadmium and zinc out of root cells into the xylem, has been isolated and purified from *T. caerulescens* roots. As the protein is naturally rich in cysteins, stability was a major problem once the protein had been purified. Therefore, all characterisation steps had to be performed immediately after purification and for each new data set, fresh protein had to be purified. Identity and puritiy have been confirmed by SDS gels and western blots. ATPase activity assays in the presence of various metals in different concentrations have been conducted. These showed that TcHMA4 is not only acitivated by zinc and cadmium, but also by copper. Nevertheless, with cadmium and zinc up to a concentration of 10μM the ATPase acitivity was increased while using 3μM of copper, the absolute phosphate concentration generated by TcHMA4 decreased slightly. This suggests that not only ATPase activity, but also ATP synthase activity can be increased by addition of copper yielding an equilibrium of hydrolysis and synthesis of ATP. As also the temperature dependence of activity has been measured, it was possible to determine the energy of activation for different metals and concentrations using Arrhenius plots. TcHMA4 did not show any changes in activation energy in the presence of different concentrations of zinc. Towards higher concentrations of copper, the activation energy increased. Performing extended x-ray absorption fine structure (EXAFS) measurements on cadmium bound to the protein, the fourier transformed data showed a peak characteristic for sulfur. This suggests that cadmium in TcHMA4 is mainly bound to cysteins and less to histidine, which is also present in the sequence and has been discussed in several articles to be involved in metal binding in the protein.

EXAFS has also been used for the analysis of copper in frozen leaf tissue of *T. caerulescens*. A very important finding was that within a population of *T. caerulescens*, a few individuals seem to be resistent to copper, while the majority of Thlaspi plants reacts very sensitively upon copper treatment. An interaction of copper with other copper atoms has been found, suggesting biomineralisation, a phenomenon that has been reported earlier for fungi. Additionally, all of our plants, especially the resistent ones, showed a high sulfur signal. The sulfur signal was most likely due to metallothioneins. This was a very interesting finding as in *T. caerulescens*, zinc and cadmium are both mainly bound by oxygen ligands and not by metallothioneins. Our finding once again shows how clearly hyperaccumulator plants can distinguish between a hyperaccumulated and a non-hyperaccumulated, probably even toxic, metal.

Zusammenfassung

In der hier vorliegenden Arbeit wurden verschiedene Aspekte der Metal-Akkumulation durch die Hyperakkumulatorpflanze *Thlaspi caerulescens* untersucht. *Thlaspi caerulescens* gehört zur Familie der Brassicaceen und hyperakkumuliert Zink. Der Ökotyp Ganges, entdeckt auf dem Gebiet einer ehemaligen Zinkmine in Südfrankreich, nimmt zusätzlich auch aktiv Cadmium auf. Aus früheren Studien ist bekannt, dass Hyperakkumulatoren Metalltransporter überexprimieren und dass der Großteil dieser Pflanzen das aufgenommene Metall in den Vakuolen von großen Epidermiszellen speichern.

Cd Akklimation und Translokation in *Thlaspi caerulescens*

Zunächst wurde die Langzeit-Reaktion von *T. caerulescens* auf Cadmium untersucht. Dazu wurden Pflanzen sechs Monate lang auf einer Nährlösung, die Cadmium beinhaltete, angezogen. Zuerst zeigten die Pflanzen Vergiftungssymptome wie eine Gelbfärbung der Blätter, wuchsen aber weiter. Nach zwei Monaten des Wachstums begannen die Pflanzen sich zu akklimatisieren und die Vergiftungssymptome verschwanden nahezu vollständig. Mithilfe von Chlorophyll-Fluoreszenz-Kinetik Messungen konnte gezeigt werden, dass während der Akklimatisierung nicht alle Zellen gleich von Toxizität betroffen waren. Die Verteilung von Cadmium in den Blättern war heterogen, manche Mesophyllzellen nahmen mehr Cadmium auf als andere. Mit der Einlagerung des Metalls in die Vakuole verschwand diese Heterogenität langsam. Desweiteren wurde entdeckt, dass Cadmium die photosynthetischen Lichtreaktionen stärker hemmt als den Calvin-Zyklus und dass es im/am Photosystem II mindestens zwei verschiedene Angriffsstellen für Cadmium gibt.

Mit Hilfe eines für Cadmium spezifischen Fluoreszenzfarbstoffes und Protoplasten aus *Thlaspi* Blättern konnte die Aufnahme von Cadmium in Mesophyll- sowie große und kleinere Epidermiszellen gezeigt werden. Die Aufnahmeraten in große Epidermiszellen waren dabei deutlich höher als die Aufnahmeraten, die für die beiden anderen Zelltypen gemessen wurden. Nachdem für die Messungen isolierte Protoplasten verwendet wurden ist es offensichtlich, dass diese Unterschiede nicht durch den Transpirationsstrom herbeigeführt werden können sondern durch eine unterschiedliche Expression von Metalltransportern in den verschiedenen Zelltypen zustande kommen.

Kurz nach der Zugabe von Cadmium zum Messmedium erschien ein heller Ring an der Innenseite der Zellen und war dort eine Weile lang zu beobachten. Erst im weiteren Verlauf wurde langsam die gesamte Zelle hell, was zeigte, dass die Translokation von Cadmium in die Vakuole ein zeitlimitierender Schritt bei der Hyperakkumulation in *T. caerulescens* ist.

Metalltransport und Detoxifizierung

Bis heute ist nicht viel bekannt über Metalltransporter in Pflanzen, insbesondere nicht über metall-transportierende ATPasen. Nachdem ATPasen eine wichtige Rolle in der Hyperakkumulation spielen, wurde TcHMA4, eine P_{1B}-Typ ATPase, die in die Metallbeladung des Xylems involviert sein soll, aus *Thlaspi*-Wurzeln aufgereinigt und charakterisiert. Weil das Protein sehr reich an Cysteinen ist, war die Instabilität des aufgereinigten Proteins hier ein großes Problem. Daher mussten sämtliche Experimente zur Charakterisierung sofort nach der Reinigung durchgeführt werden und für Wiederholungen und neue Experimente musste frisches Protein gereingt werden. Die Identität und Reinheit der jeweiligen Proteincharge wurde mittels SDS Gelen und Western Blots überprüft und es wurden ATPase-Aktivitätstests in der Gegenwart von verschiedenen Metallen in unterschiedlichen Konzentrationen durchgeführt. Diese zeigten, dass TcHMA4 nicht nur von Zink und Cadmium, sondern auch von Kupfer aktiviert wird. Auch mit der höchsten Konzentration (10µM) an Zink und Cadmium war die ATPase noch aktivierbar, während die absolute Phosphatkonzentration mit Zugabe von 3µM Kupfer bereits abnahm. Dies ist vermutlich nicht durch eine Hemmung zu erklären, sondern durch ein Gleichgewicht zwischen ATPase- und ATP-Synthase-Aktivität. Da auch die Temperaturabhängigkeit der Proteinaktivität gemessen wurde, konnte mittels Arrheniusgraphen die Aktivierungsenergie in der Gegenwart von Metall bestimmt werden. TcHMA4 zeigte in der Gegenwart von Zink keinerlei Änderungen der Aktivierungsenergie, während sie mit ansteigender Kupferkonzentration zunahm.

Mit der Anwendung von Röntgenabsorptions-Spektroskopie (EXAFS) an proteingebundenem Cadmium konnte gezeigt werden, dass Cadmium in TcHMA4 hauptsächlich von Schwefel, also von Cysteinen gebunden wird, und zu einem weitaus geringeren Anteil von Histidin. Dies war ein sehr interessanter Befund, da in der Vergangenheit immer wieder diskutiert wurde, welchen Beitrag Histidin zur Bindung von Cadmium in TcHMA4 leisten könnte.

EXAFS wurde auch für die Kupferanalyse in gefrorenen Blättern von *T. caerulescens* verwendet. Eine wichtige Entdeckung war hierbei, dass sich innerhalb einer Population einige wenige Individuen als kupferresistent erwiesen, während der Großteil der Pflanzen sehr empfindlich auf Kupferzugabe reagierte. Besonders bei den resistenten Individuen wurde eine Interaktion von Kupfer mit anderen Kupferatomen entdeckt, die auf Biomineralisation als Cu-Oxalat hinweist. Dieses Mineral wurde bereits in früheren Arbeiten an Pilzen nachgewiesen. Zusätzlich wurde in allen Pflanzen, etwas verstärkt in den resistenten Individuen, ein starkes Schwefelsignal festgestellt, welches auf Metallothionein als Kupferligand hinweist. Dies war ein weiterer interessanter Befund, da sowohl Zink als auch Cadmium in *T. caerulescens* hauptsächlich von Sauerstoffliganden gebunden werden und nicht von Metallothioneinen. Hier zeigt sich ein weiteres Mal die erstaunliche Fähigkeit von Hyperakkumulatoren, zwischen dem hyperakkumulierten Metall und nicht-akkumulierten, manchmal sogar toxischen Metallen, zu unterscheiden.

1. Introduction

1.1. State of the art

Many heavy metals such as copper, nickel and zinc are well-known as essential trace elements for cyanobacteria and plants, and even cadmium has been found to be a micronutrient for marine algae (Lane and Morel, 2000). Cadmium can occur in very high concentrations that are detrimental or even lethal to most plant species, as a result of mining (VanGeen et al. 1997), smelter activities (Buchauer 1973), dust from car tires along roadsides (Fergusson et al. 1980) or application of sewage sludge (McBride et al. 1997). High amounts of copper in the soil are found in mining areas, for example in special regions in Africa like the "copper belt" in the Republic of Kongo where copper ore naturally comes to the surface. They are also found in rivers all over Europe, especially in regions rich in vineyards, as copper is still used as an classical reagent against fungal attacks towards vine plants.

Copper is far more toxic for plants compared to animals (including humans) and bacteria because in plants, under certain conditions it can substitute the Mg^{2+} ion in chlorophyll leading to a strong decrease in photosynthetic activity (Küpper et al. 1996). Animals and bacteria do not have an accesible target for copper substitution and excess copper is excreted without doing any harm to the organism. In contrast, cadmium is far more toxic for animals and bacteria than for plants. Excess concentrations in food and/or water lead to cancer, a phenomenon not occuring in plants. For detoxification, plants use their vacuole, a large compartment in the cell with an acidic pH, as a dumping site. The vacuole only occurs in plants and in a few species of bacteria, it is not present in animal cells.

Above the threshold leading to plant growth inhibition by heavy metals, a variety of toxic effects have been observed in cyanobacteria as well as in plants. They are described in the comprehensive review on this subject by Prasad and Hagemeyer (1999) and in a more recent one by Küpper and Kroneck (2005).

In terms of heavy metals, three types of plants are existing:
- Indicator plants: they are sensitive to heavy metals and can be used as an indicator for metal in the soil, the internal heavy metal concentration is a linear function of the bioavailable heavy metal concentration in the soil (or water for aqueous plants).

- Excluders: this type of plants can tolerate heavy metals in the soil up to a special amount by preventing the accumulation of metal in the cells, either by blocking the uptake in the roots or by active (energy dependent) efflux pumps (Baker 1981).
- Hyperaccumulators: these are plants that can not only tolerate high amounts of heavy metals like zinc and cadmium, but take them up actively and accumulate them up to several percent of the dry mass of their aboveground parts. Since 1977, those plants are called hyperaccumulators (Brooks et al. 1977).

The ability of hyperaccumulators to actively take up and store metal can be used by humans for two processes: phytomining, where for example nickel is taken up by the plants which are then burned and the nickel is harvested (Li et al. 2003, Chaney et al. 2005). The second and nowadays environmentally very relevant process is the so called phytoremediation where hyperaccumulators are planted out on soil that is contaminated by a metal and therefore no agricultural plants can be grown on this particular soil. The hyperaccumulators take up the metal and can then be harvested and burned, which concentrates the metal in a much smaller volume. Especially for cadmium and zinc this process works rather well and after several growing periods many soils can be used for agricultural purposes again (McGrath et al. 2006, Maxted et al. 2007).

Hyperaccumulators take up hyperaccumulated metal(s) actively and use it to prevent attacks from herbivores and fungi (Boyd & Martens 1994; Martens & Boyd 1994; Boyd et al. 2002; Hanson et al. 2003; Jhee et al. 2005). Their ability to not only deal with high metal contents in the soil but to even use it as a defense strategy enables them to grow in niches like the soil around metal ores. It has been shown for cadmium, nickel and zinc that the main detoxification mechanism is the metal storage in the vacuoles of especially large epidermal cells (Küpper et al. 1999, 2001; Frey et al. 2000; Bhatia et al. 2004; Bidwell et al. 2004; Broadhurst et al. 2004; Cosio et al. 2005). The vacuole contains less sensitive enzymes compared to the mesophyll and therefore the metal can not cause any problems like damage to sensitive enzymes or the photosynthetic apparatus. In such "storage cells", metal concentrations in the range of up to several hundred mM can be reached (Küpper et al. 1999, 2001). The metal in storage cells has been shown to be bound mainly to weak oxygen ligands (Küpper et al. 2004, Mijovilovich et al. 2009). Until now, it is still unclear, which steps in uptake and sequestration are the time limiting ones in hyperaccumulation. Further, it is not known how the taken up metal is distributed in leaf cells directly after uptake: is it transported directly into storage cells or do mesophyll and normal sized epidermal cells act as an intermediate storage site before the metal is translocated into the final storage sites? However, it is clear that for an accumulation against a concentration gradient that high, active transport is not only necessary but also the transporters involved in the translocation process into the storage sites must be highly expressed. Starting with Pence et al. in 2000, many studies on the gene level have shown that metal transporters are indeed expressed to a much higher extent in hyperaccumulator plants compared to non-accumulating plants. This fact makes hyperaccumulator plants a natural overexpression system for metal transporters that could probably be used for the production of purified protein. As so far hardly any metal transporters from plants are characterised, mainly because homologous overexpression is causing problems when used for membrane proteins, purification from hyperaccumulators could help to circumvent some of these problems.

1.2. Aims and objectives of this thesis

The aim of this project was to investigate several aspects of metal detoxification and transport in the hyperaccumulator plant *Thlaspi caerulescens*. The ecotype Ganges (which originates from Southern France) of *Thlaspi caerulescens* hyperaccumulates both zinc and cadmium, while it is, apart from some resistant individuals (Mijovilovich et al. 2009), highly sensitive to copper.

Thlaspi caerulescens is a very good model for studying hyperaccumulation. It grows relatively fast and yields a high amount of biomass. In contrast to *Arabidopsis halleri* it is self-compatible, making seed production easy.

First, we were interested in how the plants deal with different physiologically relevant concentrations of cadmium and copper when the metal is supplied with the nutrient solution over a longer time period (months) than it was normally used in metal experiments by other authors (in most cases only several days and in very high, physiologically non-relevant concentrations). For this purpose, plants were grown for several months. By measuring chlorophyll fluorescence kinetics over the whole lifetime of the plants, we wanted to find out whether part of the Cd tolerance in *T. caerulescens* is inducible and also whether a transient physiological heterogenity might be involved as an emergency defence. This means that first also mesophyll cells might take up cadmium before the metal is translocated into storage cells. It was particularly interesting to find out wether the plants would just take up the metal until the concentration reaches a critical point leading to death or wether they probably can "acclimate" to elevated concentrations of cadmium and copper. While it is known that cadmium and zinc are mainly bound by oxygen ligands in *T. caerulescens* (Küpper et al. 2004), nothing has been reported about the ligands of copper in *T. caerulescens*. Using Extended X-ray absorption spectroscopy (EXAFS), it could theoretically be possible to determine those ligands, but rather high metal concentrations in the samples are necessary and explain why no data are available on this topic yet. Instead of using physiologically irrelevant high concentrations for yielding EXAFS data, we rather want to use low concentrations, as they naturally occur in the habitat of the plants. To still receive usable EXAFS data, the measuring time will be extended strongly, up to several days.

A further aim of this project is to study the metal transport into vacuoles of epidermal storage cells in detail using protoplasts and a fluorescent dye specific for cadmium. Although it has been known for several years that hyperaccumulated metal is stored in special storage cells, it was still not known to which extent the differences in transporter expression and/or morphologic differences are responsible for the metal translocation into storage cells. As we also did not know wether the metal is pumped directly into storage cells or first into mesophyll or normal sized epidermal cells for later distribution, it is interesting to incubate several cell types with the dye and cadmium to see if all of them or only storage cells take up the metal. Further, it was still unclear what the time limiting step(s) in metal uptake are, the translocation from root to shoot, the transport into the cytoplasm of leaf cells or the sequestration over the vacuolar membrane, out of the cytoplasm and into the final storage site, the vacuole?

Obviously, before being stored in storage cells, the metal has to be translocated from the roots over the stems to the shoot. Although many gene expression studies have been done showing various metal transporters being overexpressed, still not a single one of these transporters from plants has been isolated and purified in its native state. Until today only one heavy metal ATPase as holoenzyme, the copper transporting ATP7 (MNK) from human, has been expressed in insect cells for partial biochemical characterisation (Hung et al. 2007) and no structures of a holoenzyme are available. Many subdomains have been crystallised and their structures have been solved, but even with homologous overexpression it was not possible to produce holoprotein as stable as it is necessary for structure determination and further characterisation. We therefore want to use the natural overexpression of metal transporters for isolating a Cd/Zn ATPase, TcHMA4, from roots of *T. caerulescens*. As membrane proteins are difficult to isolate and even more, to purify, a method for this purpose has to be developed first. After obtaining purified protein, characterisation via various assays yields insights into the function of TcHMA4. Using an ATPase activity test, activation patterns by metals are an interesting topic to study. Even more so, with the use of an temperature gradient in addition to different metal concentrations, the activation energy of TcHMA4 could be calculated using Arrhenius plots. Further, EXAFS is an appropriate method for searching for metal ligands in the protein, unfortunately a high amount of pure protein is a prerequisite for successful measurements using this technique.

Thus, from whole plant physiology to cellbiology and proteinbiochemistry several aspects of metal transport are investigated.

2. Publications in peer-reviewed journals and manuscripts

2.1. Cadmium-induced inhibition of photosynthesis and long-term acclimation to cadmium stress in the Cd hyperaccumulator *Thlaspi caerulescens*

Hendrik Küpper[1,2*], Aravind Parameswaran[1], Barbara Leitenmaier[1], Martin Trtílek[3], and Ivan Šetlík[2,4]

1) Universität Konstanz, Mathematisch-Naturwissenschaftliche Sektion, Fachbereich Biologie, D-78457 Konstanz, Germany
2) Faculty of Biological Sciences and Institute of Physical Biology, University of South Bohemia, Branišovská 31, CZ-370 05 České Budejovice, Czech Republic
3) Faculty of Biological Sciences and Institute of Physical Biology., Koláčkova 31, CZ-62100 Brno, Czech Republic
4) Microbiological Institute, ASCR, Department of Autotrophic Microorganisms, Opatovický mlýn, CZ-37981 Třeboň, Czech Republic

published in 2007 in New Phytologist 175: 655-674

SUMMARY

Acclimation of hyperaccumulators to heavy metal-induced stress is crucial for phytoremediation and was investigated using the hyperaccumulator *Thlaspi caerulescens* and the non-accumulators *T. fendleri* and *T. ochroleucum*.

The main technique was the measurement of spatially and spectrally resolved kinetics of in vivo absorbance and fluorescence with a novel fluorescence kinetic microscope.

Key Results: While at the beginning of growth on cadmium all species suffered from toxicity, T. caerulescens completely recovered later. During stress, a few mesophyll cells in *T. caerulescens* became more inhibited and accumulated more Cd than the majority. This heterogeneity disappeared during acclimation. Chlorophyll fluorescence parameters related to photochemistry were more strongly affected by Cd-stress than non-photochemical parameters, and only photochemistry showed acclimation. In healthy plants, maximal PSII efficiency (Fv/Fm) was homogeneous from 650-790 nm. Cd-stress reduced Fv/Fm and ΦPSII stronger at 670-740nm compared to >740nm and <670nm. Cd enhanced nonphotochemical quenching mainly <690nm.

Main Conclusions: Cd-acclimation in the Cd-hyperaccumulator *T. caerulescens* shows that part of its Cd tolerance is inducible and involves transient physiological heterogeneity as an emergency defence mechanism. Differential effects of Cd-stress on photochemical vs. non-photochemical parameters indicate that Cd inhibits the photosynthetic light reactions more than the Calvin-Benson cycle. Differential spectral distribution of Cd-effects on photochemical vs. non-photochemical quenching shows that Cd inhibits at least two different targets in/around PSII. Spectrally homogeneous Fv/Fm suggests that in healthy *T. caerulescens* all chlorophylls fluorescing at room temperature are PSII-associated.

Key words: Acclimation, Cadmium, Heterogeneity, Imaging and spectral measurements of chlorophyll fluorescence kinetics, Metal sequestration, Photosynthetic performance

Abbreviations:

F_0 = minimal fluorescence yield of a dark-adapted sample, fluorescence in non-actinic measuring light

F_m = maximum fluorescence yield of a dark-adapted sample

F_s = steady state fluorescence under the given actinic irradiance, i.e. after the end of the induction transient

F_v = variable fluorescence; $F_v = F_m - F_0$

F_p = fluorescence yield at the P level of the induction curve after the onset of actinic light exposure

LHC = light harvesting complex

Mg-substitution = substitution of the natural central ion of Chl, Mg^{2+}, by heavy metals

NPQ = non-photochemical quenching, in this manuscript used as an acronym for the name of this phenomenon. In this manuscript, we measure non-photochemical quenching as $q_{CN} = (F_m - F_m')/F_m$ = "complete non-photochemical quenching of Chl fluorescence", i.e. with normalisation to F_m.

OD = optical density

PSII = photosystem II

RC = photosynthetic reaction centre

$\Phi_{PSII} = \Phi_e = (F_m' - F_t')/F_m'$ = effective quantum yield of photochemical energy conversion in actinic light (Genty et al. 1989). In the current manuscript the values of this parameter were calculated also for responses to saturating flashes during the relaxation period after the end of actinic light in order to follow the return of the system to its dark-acclimated state as measured by F_v/F_m.

INTRODUCTION

Heavy metals such as copper, manganese, nickel and zinc are well known to be essential microelements for the life of plants, and even cadmium has been found to be the natural active site of an enzyme (Lane & Morel, 2000). However, elevated concentrations of these metals induce inhibition of various processes in plant metabolism (reviewed e.g. by Prasad & Hagemeyer, 1999; Joshi & Mohanty, 2004; Küpper & Kroneck, 2005). Cadmium can occur in the environment in high concentrations as a result of various human activities (Lagerwerff & Specht, 1970; Buchauer, 1973; Fergusson et al. 1980; McBride et al. 1997; VanGeen et al. 1997). Photosynthetic reactions belong to the most important sites of inhibition by many heavy metals, including Cd, under environmentally relevant concentrations. In the thylakoids, photosystem II (PSII) has frequently been identified to be the main target. The type of damage, however, strongly depends on the irradiance conditions (Cedeno-Maldonado, 1972; Küpper et al. 1996, 1998, 2002, 2006). The latter authors found that in low irradiance including a dark phase the inhibition of PSII is largely due to the impairment of the correct function of the light harvesting antenna, this mechanism was termed "shade reaction". It results from the substitution by heavy metals of the Mg^{2+} ion in the chlorophyll molecules of the LHCII. In high irradiance, direct damage to the PSII reaction centre (PSII RC) occurs instead; this was named "sun reaction" (Küpper et al. 1996, 1998, 2002). A recent study suggested competitive binding of Cd^{2+} to the essential Ca^{2+} binding site (Faller et al. 2005). But this was only tested in isolated PSII RC particles; its relevance in vivo is unknown.

Plants developed many strategies to resist the toxicity of heavy metals as reviewed e.g. by Prasad & Hagemeyer (1999), Cobbett & Goldsbrough (2002) and Küpper & Kroneck (2005). Most heavy metal tolerant plants prevent accumulation of heavy metals in their above-ground tissues (i.e. have a bioaccumulation coefficient < 1) and are therefore called "excluders" (Baker, 1981). This exclusion can be achieved in several ways, as reviewed e.g. by Küpper and Kroneck (2005). It involves reduced translocation from the root to the shoot, but also true exclusion from the roots, e.g. via lignification (Cuypers et al. 2002) and ATP-dependent efflux pumps (van Hoof et al. 2001). Other metal-tolerant plants actively take up heavy metals, translocate them into the shoot and sequester them to certain parts of the plant, where they are stored in a harmless state. These plants, called "hyperaccumulators", accumulate up to several percent heavy metal in the dry mass of their aboveground parts (Brooks et al. 1977). In their natural habitats this hyperaccumulation most likely serves as a defence against pathogens and herbivores, as demonstrated by many studies (Boyd & Martens, 1994; Martens & Boyd, 1994; Boyd et al. 2002; Hanson et al. 2003; Jhee et al. 2005). However, certain herbivores and pathogens attack hyperaccumulators despite their high content of toxic metals, so that some authors dispute that adaptive value of hyperaccumulation (reviewed by Küpper & Kroneck 2005; Poschenrieder et al. 2006). Hyperaccumulators can be used for the decontamination ("phytoremediation") of anthropogenically heavy metal contaminated soils, and in some cases also for the commercial extraction ("phytomining") of high value metals (mainly Ni) from metal-rich soils (e.g. Baker et al. 1994; McGrath & Zhao, 2003; Chaney et al. 2005). The best known Zn-hyperaccumulator species is *Thlaspi caerulescens* J.&C. Presl., which has been proposed as a hyperaccumulator model species by several authors (Assunção et al. 2003; Peer et al. 2003, 2006). The "Ganges" ecotype (= "French A" in Lombi et al. 2000) of this species was also the first plant for which real Cd hyperaccumulation in the sense of > 1% Cd in the shoot dry mass has been shown.

The mechanisms by which plants hyperaccumulate heavy metals in their shoots, and prevent phytotoxicity of these metals, have been a subject of many studies. Nevertheless, many of these mechanisms are still under debate (e.g. Pollard et al. 2002; Küpper & Kroneck 2005). While hyperaccumulators require mechanisms of metal detoxification to allow the plants to survive, metal hyperaccumulation and resistance are genetically independent characters (Zn: Macnair et al. 1999; Cd: Bert et al. 2003). An enhanced uptake of Zn into the root symplasm was found in the Zn/Cd-hyperaccumulator *T. caerulescens* compared to the related non-accumulator *T. arvense* (Lasat et al. 1996, 1998), and a reduced sequestration into the root vacuoles was associated with the higher root to shoot translocation efficiency of *T. caerulescens* (Shen et al. 1997; Lasat et al. 1998). In most hyperaccumulators the metal (shown so far for Cd, Ni, Zn) is sequestered preferentially into compartments (usually the epidermal vacuoles) where it does least harm to the metabolism (Küpper et al. 1999, 2001; Frey et al. 2000, Bidwell et al. 2004; Bhatia et al. 2004; Broadhurst et al. 2004; Cosio et al. 2005). The approximate volume of this storage site multiplied by the metal concentration in it (data e.g. for Zn from Küpper et al. 1999) indicates that about 70% of the total accumulated metal in mature leaves is stored in the epidermis. In the vacuoles of the main epidermal metal storage cells, concentrations of several hundred mmol.l^{-1} can be reached (Küpper et al. 1999, 2001), showing that hyperaccumulation must be mediated by active pumping of the metals into their storage sites. Indeed, starting with Pence *et al.* (2000), many studies have shown that metal hyperaccumulation is caused by an extremely increased expression of metal transport proteins compared to non-accumulator plants (Assunção et al. 2001; Becher et al. 2004; Papoyan & Kochian, 2004; Weber et al. 2004). Strong sulphur ligands like phytochelatins were shown not to be relevant for cadmium detoxification in the Cd hyperaccumulator *T. caerulescens* by several authors (Ebbs et al. 2002; Schat et al. 2002; Küpper et al. 2004), and the same was found for Ni (Krämer et al. 1996; Sagner et al. 1998), Zn (e.g. Salt et al. 1999; Küpper et al. 1999, 2004) and As (Wang et al. 2002; Webb et al. 2003). So the main detoxification strategy in hyperaccumulators for hyperaccumulated metals (Cd, Ni, Zn and the metalloid arsenic) is clearly the sequestration of these metals.

However, even hyperaccumulators have a limited tolerance towards heavy metals. Under conditions of excess supply of the hyperaccumulated metal, metal accumulation was found to be enhanced in a few cells of the mesophyll (Cd: Küpper et al. 2000a; Ni: Küpper et al. 2001). These same cells were found to contain elevated levels of magnesium, which was interpreted as a defence against substitution of Mg^{2+} in Chl (Küpper et al. 1996, 1998, 2002) by heavy metals. It remained unknown, however, in which way this metal accumulation heterogeneity under stress is coupled to heterogeneity in physiology, e.g. mesophyll photosynthesis. But so far it has never been studied in detail how photosynthesis changes in hyperaccumulators during heavy metal-induced stress and acclimation to it. These processes are important in terms of using hyperaccumulators for the phytoremediation or phytomining of metal-rich soils, because they set the frame that limits the biotechnological use of these plants. It has to be known, for example, within which time and up to which metal level acclimation efficiently occurs, and which physiological parameters are reliable indicators to decide whether the plants will manage to acclimate or rather die if no action is taken by the farmer. Knowledge of the acclimation response is particularly important for judging about the development of a crop of hyperaccumulators on highly contaminated soils, and on soils with a leached-out top layer. In the first case, the hyperaccumulators would already germinate in high metal concentrations, and would have to acclimate to heavy metal toxicity already at the beginning of growth. In the second case, hyperaccumulators would germinate in low-metal conditions and become exposed to toxic concentrations when the roots grow into the high-metal deeper soil. Leached-out uppermost few centimetres of soil are usual in naturally Cd-rich habitats (e.g. McBride et al. 2005) and in anthropogenically contaminated sites where heavy metal input from the atmosphere or by sewage sludge stopped a long time ago (e.g. Mitani & Ogawa, 1998), so we simulated this situation in the study presented here.

Measurement of chlorophyll fluorescence kinetics is a powerful method for investigating the physiological status of plants (recently reviewed e.g. by Maxwell & Johnson, 2000, Roháček, 2002; Papageorgiou & Govindjee, 2004), in particular in cases like heavy metal-induced stress where photosynthesis is a primary target of inhibition (reviewed by Joshi & Mohanty, 2004; Küpper & Kroneck, 2005). Chlorophyll fluorescence kinetic measurement is a more direct way to assess photosynthetic activity than the old popular method of gas exchange measurement, which only measures the difference between photosynthetic oxygen release (or carbon dioxide uptake) and oxygen consumption (or carbon dioxide release) by various respiratory processes. Both the power and popularity of Chl fluorescence kinetic measurements were further increased by the introduction of spatially resolved (=two-dimensional, imaging) measurements with calibrated CCD cameras (reviewed by Nedbal & Whitmarsh, 2006, and Oxborough, 2004). Such measurements can directly investigate differences between photosynthetic entities, such as chloroplasts, cells, tissues, organs or individuals. A macroscopic imaging system using the pulse amplitude modulation principle of measurement was developed by Nedbal et al. (2000) and a microscope using the same principle, the "Fluorescence Kinetic Microscope" (FKM), was developed by Küpper et al. (2000b). Chlorophyll fluorescence imaging made the spatial heterogeneity of photosynthetic processes a commonly known fact (Siebke & Weis, 1995a, 1995b; Nedbal et al. 2000; Küpper 2000b; Ferimazova et al. 2002). The static fluorescence pictures of Küpper et al. (1998) showed that in *Elodea canadensis* the effect of exposure to Cu^{2+} might strongly differ in adjacent groups of cells, and macroscopic Chl fluorescence imaging of Cu-stressed tobacco leaves showed heterogeneity as well (Ciscato & Valcke, 1998). Moreover, x-ray emission spectroscopy on frozen-hydrated leaves showed that heavy metal distribution in the leaf mesophyll cells might be rather heterogeneous (Küpper et al. 2000a, 2001). These findings made it tempting to investigate the effects of Cd^{2+} on the photosynthetic characteristics in the mesophyll of treated plants in more detail by microscopic fluorescence imaging.

In the current study, we investigated cadmium-induced inhibition of photosynthesis and long-term acclimation to this stress in the Cd/Zn-hyperaccumulator model plant (Assunção et al. 2003; Peer et al. 2003, 2006) *T. caerulescens* compared to the closely related Cd/Zn non-accumulators *T. fendleri* and *T. ochroleucum*. This was done by continuously monitoring the state of the plants in long-term (up to 12 months) experiments with Cd concentrations occurring in the habitat of *T. caerulescens*. We focussed on the biophysical aspects of the processes, because in contrast to the molecular biological aspects (such as metal transporter expression) so far the former have been much less investigated in hyperaccumulators. The most important method for our investigations was the imaging and spectrally resolved microscopic measurement of chlorophyll fluorescence kinetics under physiological conditions. This was done with a new version of the fluorescence kinetics microscope (FKM) built for this study, illustrated in Fig. 1 and described in the methods. The new FKM differs from the previous one (Küpper et al. 2000b) mainly by providing spectrally resolved fluorescence kinetics of one selected area in the picture in addition to the spatially resolved (imaging) fluorescence kinetics. This was now used to identify the inactivation of specific parts of the photosynthetic antenna by Cd. Further, easily selectable spectral bands for excitation allow measurement of chlorophyll fluorescence kinetics with excitation specific for various parts of the light harvesting antenna, as well as measurement of any kind of non-chlorophyll fluorescence kinetics. In this paper, the latter was used to monitor Cd accumulation via a specific fluorescent dye. The FKM measurements were performed on the background of classical physiological measurements of growth and pigment content.

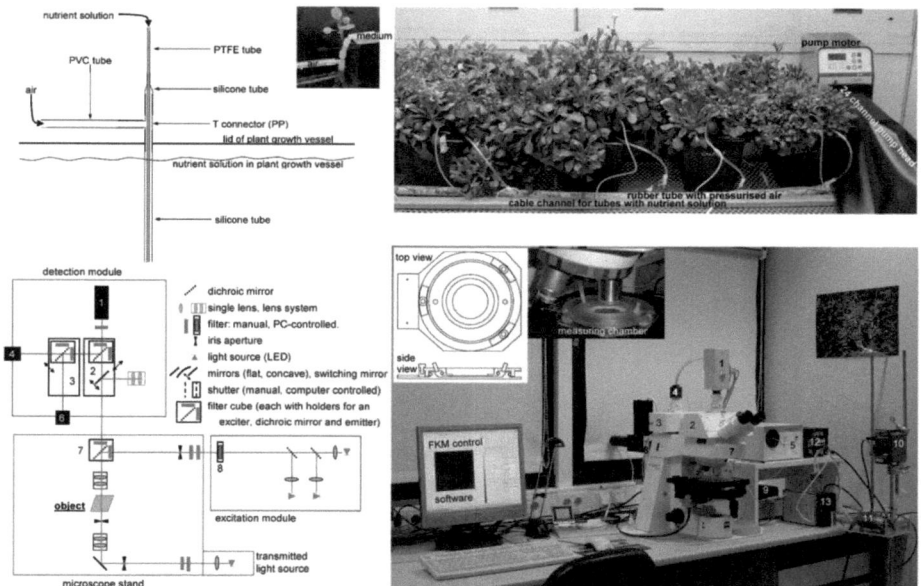

Fig. 1. New experimental technology used in the second block of experiments in this study, as describd in more detail in the methods. Top row: the hydroponic growth system. Left: scheme of a media injector. They were constructed in a way that each drop of fresh nutrient solution coming out of the tube is immediately thoroughly mixed with the medium in the culture vessel by the air. Middle: Photograph of a media injector installed in the system. Right: Photograph of the system including the pump, the supply tubes, and a few of the plants. Bottom row: the new version of the Fluorescence Kinetic Microscope (FKM). Explanation of numbers on the scheme and photograph. 1: measuring camera, 2: 3-port motorised video adapter, 3: 2-port manual switching adapter, 4: fibreoptics adapter for spectrometer, 5: spectrometer, 6: C-mount adapter for photo camera, 7: motorised wheel for filter cubes (8 cubes) , 8: filter wheel for switching excitation while maintaining emission filters, 9: FKM control unit, 10: peristaltic pump for liquid media, 11: air pump, 12: temperature control unit, 13: flow-through thermostat Left: Scheme of the optical system. Right: Photograph of the complete FKM. Inset: close-up view of the measuring chamber as a photograph and a scheme.

MATERIALS AND METHODS

Plant material, culture media and culture conditions

Seeds of *Thlaspi caerulescens* J.&C. PRESL (Ganges population), *Thlaspi fendleri* (NELS.) HITCHC (= *Thlaspi montanum* ssp. montanum according to Koch and Al-Shehbaz, 2006, i.e. it may be a nickel hyperaccumulator) and *Thlaspi ochroleucum* BOISS ET HELDER were germinated on a 3:1 perlite:vermiculite mixture moistened with deionised water. Three weeks after germinatiön, seedlings were transferred to a nutrient solution containing 1000 µM $Ca(NO_3)_2$, 500 µM $MgSO_4$,

50 µM K_2HPO_4, 100 µM KCl, 10 µM H_3BO_3, 0.1 µM $MnSO_4$, 0.2 µM Na_2MoO_4, 0.1 µM $CuSO_4$, 0.5 µM $NiSO_4$, 20 µM Fe(III)- EDDHA (Fe(III)-ethylenediamine-di(o-hydroxyphenylacetic acid), and 10 µM $ZnSO_4$ (i.e. like Shen *et al.* 1997, but lower Cu and Mn). The pH of the solution was maintained at around 6.0 with 2000 µM MES-KOH (2-morpholinoethanesulphonic acid). As in many previous studies (e.g. Küpper et al. 1999, Lombi et al. 2000) the nutrient solution contained high (10 µM) Zn because of the high Zn requirement of Zn hyperaccumulators (e.g. Shen et al. 1997). The nutrient solution was aerated continuously. All chemicals were analytical grade and purchased from Merck (Darmstadt, Germany; www.merck.de) except for Fe-EDDHA, which was purchased from Duchefa Biochemie (Haarlem, The Netherlands; www.duchefa.com).

This study altogether lasted for seven years, and the experiments were carried out in two blocks with the development of a new FKM (see below) in between. Altogether, for the data presented here, stress+acclimation were investigated in 9 experiments, including 55 pots with 280 plants of *T. caerulescens*, of those 25 pots with 108 plants were Cd-treated. For *T. fendleri* we analysed 10 pots with 38 plants and for *T. ochroleucum* 6 pots with 24 plants. In our first series of four experiments (all of them with *T. caerulescens* and one with all three species, 2000-2001), we used 1.5 l vessels with 3-4 plants each and renewed the solutions manually every 4 d (i.e. renewal rate per plant about 85 ml.d^{-1}). In the second series (five experiments, all of them with *T. caerulescens* and two with *T. fendleri*, 2004-2007), we used 6 l vessels with 7 plants each, in which the solution was exchanged continuously (1700 ml.d^{-1} per pot, i.e. 250 ml.d^{-1} per plant) with the programmable 24-channel peristaltic pump "MCP Process" (Ismatec, Glattbrugg, Switzerland, www.ismatec.com). The solutions in the pots were constantly thoroughly mixed via a lab-built media injection system (Fig. 1). Two weeks after transferring the seedlings into hydroponic solution, Cd^{2+} was added as specified in table 1. The increased flow rate in the second experiment series was chosen to make sure that Cd uptake into the plants was not limited by the total amount available in the solution, but only by the concentration. For *T. caerulescens* but not for *T. fendleri* and *T. ochroleucum*, this yielded stronger stress compared to the earlier experiments, showing that there the response of *T. caerulescens* was limited by the total cadmium per pot due to hyperaccumulation in the plants.

All plants were grown with 14 h day length. In the first series of experiments, 24 °C/20 °C day/night temperature, and a constant irradiance of 60 µE (from a 1:1 mixture of "cool white" and Fluora® fluorescent tubes, OSRAM, München, Germany, www.osram.com) during the light period was applied. In the second series, 22°C/18°C day/night temperature was applied and a quasi-sinusoidal 3-step light cycle with about 40 µE in the morning and 120 µE at noon was achieved by full spectrum discharge lamps.

Despite the differences in growth conditions between the first and the second series of experiments, all trends of changes in photosynthetic parameters and growth were found in both series, so that all experiments will be analysed together in the results and discussion.

The new version of the Fluorescence Kinetic Microscope

For the study presented here, the most important method for analysing physiological performance of the plants was the two-dimensional (imaging) microscopic measurement of chlorophyll fluorescence kinetics, with which photosynthetic performance can be assessed on a single-cell level. The first instrument that could perform this task via the pulse amplitude modulation principle of fluorescence measurement in a microscope, the "Fluorescence Kinetic Microscope" (FKM, Küpper et al. (2000b), was used for the 2000-2001 series of experiments. The new version of the FKM (Fig. 1) was constructed 2004 in collaboration with Photon Systems Instruments (PSI: Brno, Czech Republic,www.psi.cz) and was used for the 2004-2007 series of experiments. It differs from the previous version mainly in the following. (1) All excitation lights are provided by high-output white LEDs with a software-controlled timing and intensity. From the spectrum of the LEDs the excitation bands are selected by filters mounted in motorised wheels that are controlled by commands from the software, similar to filters controlling the emission spectrum. (2) The major advancement is the option to use a newly developed high-sensitivity fibre optic spectrometer (based on the module MCS-CCD from Carl Zeiss, Jena, Germany) sensitive enough to record spectra of F_0 in non-actinic measuring light, and to record the Kautsky kinetics of a single cell with 10 ms time resolution. The light coming from the sample is divided as shown in Fig. 1 and can be directed either 100% to the camera or divided between camera and spectrometer (usually 90% were sent to the spectrometer for spectrally resolved kinetics). The spectrometer is synchronised to the camera and can be operated through the FluorCam6 software, so that both spatially and spectrally resolved fluorescence kinetics can be recorded simultaneously. (3) The camera is the same as in the commercially available FluorCam systems (PSI, Brno, Czech Republic). It can capture 50 frames.s^{-1} at down to 10 µs exposure time per frame, 70% peak quantum yield, 4 electrons readout noise, 12 bit dynamic range and 512x512 pixels spatial resolution. (4) It is based on an Axioplan2 imaging microscope (Zeiss, Jena, Germany, www.zeiss.com). This model offers many new possibilities that were crucial for the present work, in particular the control most of its functions by the FluorCam6 software.

The FKMs were controlled, and the measurements analysed (see below), by the commercial FluorCam software (version 5 for the earlier, version 6 for the newer series of experiments) from Photon Systems Instruments (Brno, Czech Republic, www.psi.cz) that was, however, developed with the additional demands of the FKM hardware in mind.

Plant species	Cd^{2+} conc / µM	plant fresh weight / g (± SE)			shoot/root fresh weight ratio (± SE) at time of harvest	Pigment content in shoot at the time of harvest /µmol (g dry weight)$^{-1}$ (± SE)		Chl a / Chl b ratio (± SE) at the time of harvest
		after 2 months	after 5 months	after 7.5 months		Chl a	Chl b	
T. caerulescens	0	5.80 (1.76)* n=5	52.4 (4.2) n=17	158.4 (10)# n=7	1.76 (0.17) n=7	5.55 (0.05) n=7	1.96 (0.01) n=7	2.83 (0.00) n=7
	10	-	15.1 (6.4) n=4	52.5 (8.4)# n=19	1.60 (0.30) n=19	3.24 (0.34) n=12	1.19 (0.15) n=12	2.78 (0.06) n=12
	50	1.08 (0.18)*$ n=4	1.99 (0.47) n=12	0.70 (0.14)# n=7	1.1 (0.23)	0.27 (0.04) n=7	0.09 (0.01) n=7	2.86 (0.05) n=7
T. fendleri	0	2.66 (1.06)* n=4	58.2 (6.9)# n=10	-	0.48 (0.18) n=10	5.99 (0.26) n=12	2.11 (0.12) n=12	2.86 (0.04) n=12
	10	0.71 (0.34)* n=4	0.41 (0.12)# n=8	-	1.73 (0.32) n=8	0.27 (0.05) n=8	0.13 (0.05) n=8	2.45 (0.28) n=8
	50	0.065 (0.026)* n=4	all plants dead	-	-	-	-	-
T. ochroleucum	0	0.92 (0.13)*# n=4	-	-	-	-	-	-
	10	1.13 (0.16)*# n=4	-	-	-	-	-	-
	50	0.045 (0.019)*# n=4	-	-	-	-	-	-

Table 1. Metal treatments and their effect on growth and pigment content of the plants. The metal treatments were set up with respect to the fact Thlaspi caerulescens is a Cd/Zn-hyperaccumulator plant (Küpper et al. 1999, 2000), while Thlaspi fendleri and Thlaspi ochroleucum are non-accumulator plants. The data are means and standard errors from three experiments of the newer (2004-2007) series unless stated otherwise and are a typical example of the behaviour of the plants as it was consistently observed and documented in all nine experiments (altogether 280 plants of T. caerulescens, 38 of T. fendleri and 24 of T. ochroleucum, see methods for details on statistics). n = number of plants analysed. - = not measured, # = all remaining plants harvested after this measurement and subjected to pigment analysis * = data from earlier series of experiments; $ = toxicity limited by flow rate of the nutrient solution (see methods).

Imaging chlorophyll fluorescence kinetic measurements

Macroscopic measurements. The FluorCam instrument (closed version; Photon Systems Instruments, www.psi.cz) described by Nedbal *et al.* (2000) was used for recording images of fluorescence kinetics (300x400 pixels at 8 bit greyscale each). All macroscopic measurements were done with blue actinic light (LEDs with 465 nm peak). A whole plant or leaf was kept in the water-saturated air of the closed FluorCam at about 20°C, and supplied with water through the roots or leaf petiole.

Microscopic measurements. Blue (410-500 nm) excitation was used, provided by white LEDs with the excitation filter 2P-HQ 460/80 (AHF, Germany, www.ahf.de) and dichroic mirror 505DCXR (AHF). Chl fluorescence was detected from 665-705 nm with the emitter filter D680/30 (AHF). Measuring light was less than 1 µE (and <0.5 µE in the 2005-2007 series of experiments).

To perform a measurement, a leaf was cut off and pressed by its upper side (for palisade mesophyll measurements) or lower side (for spongy mesophyll measurements) towards the glass window of the measuring chamber with a wet nylon grid or wet cellophane. The chamber was ventilated by a stream of water-saturated air (2 l.min^{-1}, 21°C). The construction and operation of the chamber is in principle described in Küpper *et al.* (2000b); the new version is shown in inset of Fig. 1.

All microscopic measurements were done on the mesophyll away from the veins. They lasted 300 s, and typical records are shown in Fig. 8. In the third second a 600 ms flash of saturating light was given for measurement of F_m. This was followed by 90 s of darkness, after which F_0 was measured for 5 s. Then, 100 s of actinic light were applied to analyse the Kautsky induction, and finally 100 s of measurement with no actinic light were used to measure dark relaxation and F_0'. During the actinic light exposure and in the relaxation period, 600 ms saturating flashes were applied for analysis of photochemical and non-photochemical quenching.

Microscopic *in vivo* VIS spectroscopy

Cells of which absorbance and/or fluorescence spectra should be recorded were selected according to previous imaging records of Chl fluorescence kinetics (see above), so that both types of information would be available for the same cells. The radiation to be spectrally analysed is brought to the spectrometer by a light guide, the collecting end of which is situated in the centre of a twin field of view identical to that seen by the camera (see description of the new FKM above) and has a diameter of about 10% of the width of that field. The light path was set up in a way that the spectrometer receives the light before the emitter filter that selects the fluorescence for the camera. In this way, part of the excitation light reflected by the sample in the region below 500 nm reached the spectrometer and could be used as an internal standard for calculating quantum yields. This desired crosstalk signal did not interfere with the detection of Chl fluorescence, because the excitation filter applied (2P-HQ 460/80 from AHF, Germany) was blocked with OD6 light from 510-1100 nm. The recording of the spectra was performed with the software SpectraWin (Photon Systems Instruments, see above) that allowed for a synchronisation with the FluorCam software controlling the camera and the light sources. All further analyses were performed in Microcal Origin 7.0/7.5 (Northampton, USA; www.originlab.com) and Microsoft Excel 97/2003 (Redmont, WA, USA, www.microsoft.com).

Static measurements of single-cell emission and absorption spectra. Non-kinetic single-cell emission spectra were measured with high intensity (50 µE) measuring light. Blue (410-500 nm) excitation was used as for the imaging records. The light source for absorption spectra was the white LED used also as transmittant light in the FKM (Fig. 1). To reduce noise in the spectra, the LED intensity was adjusted so that the peak intensity saturated about 95% of the dynamic range (where it was still linear) of the CCD detector of the spectrometer. Further noise reduction was achieved by averaging. This was already enough for recording high quality spectra up to OD 6. Further stretching of the dynamic range up to OD 12, which was necessary for recording the absorbance of thick leaves (Fig. 10), was achieved by recording the sample spectrum (I) of very dense samples with a higher light intensity than the reference spectrum (I_0), and subsequent recalculation of the true reference by the known relation between the intensities.

Spectrally resolved single-cell fluorescence kinetics. These measurements were carried out using the FluorCam6 software like for the regular spatially resolved measurements. The protocol defining the timing of measuring, actinic and saturating light was in principle the same as well, i.e. following the description above, except for the following. (a) It included commands to control the spectrometer hardware in addition to the measuring camera. (b) In order to have enough light for the spectra at high time resolution, the duration of the measuring light flashes was increased to 100 µs. Therefore, the measuring light during F_0 measurement was about 2 µE, but the actinic effect was still negligible.

Analysis of fluorescence kinetics

Spatially resolved (imaging) measurements. These data were analysed using the FluorCam software from Photon Systems Instruments (Brno, Czech Republic) as described earlier (Küpper et al. 2000b; Ferimazova et al. 2002). The heterogeneity visible on images of fluorescence kinetic parameters (e.g. F_m, F_0, F_v, F_v/F_m, Φ_{PSII}) was used to select objects for further analysis. The selected objects were passed again through the analysis routine, which resulted in fluorescence kinetic traces representing the average of the kinetics of all pixels within the chosen objects. These kinetic traces were exported to ASCII files for further analysis. The cells used for statistical analysis (incl. Figs. 3, 8, 9 and Supplement-Fig. 1), were selected to be a representative subset of all cells in the field of view. That means, for example, if half of the cells in the field of view displayed reduced F_v/F_m and the other half did not, 5 cells of each half were selected. To verify correct selection, the fluorescence parameter values of the selected cells were averaged and this average value was compared to the corresponding average of the whole field of view.

Spectrally resolved measurements.
In the spectrally resolved records, the measuring camera was operated in the same way as in the usual spatially resolved records, so that these measurements could be analysed as described above. For the simultaneously recorded spectrally resolved kinetics, in contrast, not only the fluorescence response to the excitation by measuring flashes, but also the fluorescence excited by the actinic light and the saturating flashes was recorded by the spectrometer. Since the latter responses are much stronger than the response to measuring light, this offered the advantage of reduced noise for the measurement of all fluorescence parameters that are recorded during actinic or saturating light. But at the same time direct assessment of the quantum yield offered by the constant power of the measuring flashes was lost. In order to extract the relative quantum yields from such records, the excitation light reflected by the sample that reached the spectrometer (see section Static measurements of single-cell emission and absorption spectra) was used as an internal standard for calculation. The raw spectrally resolved kinetic datasets, consisting of time-encoded series of spectra, were loaded into a self-designed add-in program for Microsoft Excel. Using the reflected light peaks, it automatically detected the light level (dark noise, measuring, actinic and saturating light), subtracted the dark noise, normalised the fluorescence signal to the reflected light peaks (=internal standard, see above) and finally extracted the fluorescence parameters (e.g. F_v/F_m, Φ_{PSII}, NPQ).

Quantification of chlorophylls in plant extracts.
Spectra of pigment extracts were measured with the UV/VIS spectrophotometer Lambda 16 (Perkin-Elmer, Germany) at a spectral bandwidth of 1 nm with 0.2 nm sampling interval from 350-750 nm. Chls were quantified according to the "Gauss Peak Spectra" method of Küpper et al. (2000c). Cd-Chl could not be estimated due to its spectral similarity to Mg-Chl in the 550-750 nm spectral range (Küpper et al. 1996).

Qualitative visualisation of Cd distribution in living tissue.

Accumulation of Cd in the mesophyll was visualised by the cell-permeable fluorescent dye Rhod-5N (Invitrogen, www.invitrogen.com). This dye was originally developed for calcium visualisation, but its binding and fluorescence response is 2-3 orders of magnitude higher for Cd^{2+} than for Ca^{2+}. Therefore, at the Cd^{2+} concentrations found in hyperaccumulators (e.g. Küpper et al. 1999, Cosio et al. 2005) Rhod-5N fluorescence will almost entirely reflect the (labile bound, Küpper et al. 2004) Cd^{2+} and not the Ca^{2+} concentration. To visualise Cd^{2+}, the lower epidermis of a leaf was stripped off carefully with micro-forceps. Afterwards, leaf pieces of about 3x3 mm were excised, vacuum infiltrated with infiltration medium (IM = 1 M sorbitol, 10 mM MES adjusted to pH 5.5 with BTP) containing 100 µM of Rhod-5N acetomethyl ester, and incubated in this solution at room temperature for 60 min. Afterwards, the tissue was washed twice with IM without Rhod-5N, and Rhod-5N fluorescence in the cells was measured with 530-560 nm excitation (AHF HQ 545/30), 565 nm dichroic mirror (AHF 565 DCLP) and 575-630 nm cut-off filter (AHF D605/55; all filters from AHF, Germany, www.ahf.de).

RESULTS

Growth, metal uptake, visible symptoms of damage and pigment content

This study included a sublethal but still toxic concentration of cadmium (10 µM) that is commonly found in heavy metal rich habitats, and a strongly toxic Cd concentration (50 µM). Both were applied for several months until the plants either acclimated or died. As expected, in the long run the Cd hyperaccumulator *T. caerulescens* was much more resistant than the Cd non-accumulators *T. fendleri* and *T. ochroleucum* (see fresh weights in Table 1). After five months of growth with 10 µM Cd, for example, *T. caerulescens* reached about 30% of the control fresh weight, in contrast to only 0.7% for *T. fendleri*. The metal uptake was in the usual range for these plants, with up to 2% Cd in mature leaves of *T. caerulescens* grown on 50 µM Cd^2 as documented in our publication on cadmium ligands in *T. caerulescens* (Küpper et al. 2004), in which some plants of the present study were used.

Under cadmium stress, in all species chlorotic areas appeared on leaves growing during the Cd treatment, while leaves that were mature before the onset of Cd treatment usually did not show this damage symptom. The chlorotic areas were mainly mesophyll areas distant from veins, while the mesophyll close to veins remained normally green. This symptom gradually disappeared during acclimation in *T. caerulescens*; >6 months old plants after successful acclimation to Cd looked much healthier, almost like control plants (Fig. 2). Note that this change in visible phenotype occurred much later than the physiological change measured by the fluorescence kinetics described below. Acclimation was also evident in the pigment content at the time of harvest, which in fully acclimated plants was very close to the control (Table 1). The Chl a / Chl b ratio remained almost unaffected during the whole experiment.

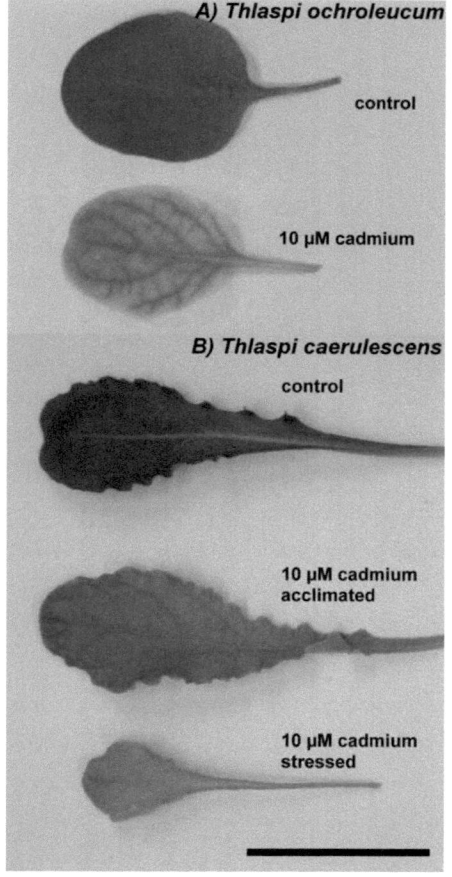

Fig. 2. Close-up view of cadmium-induced damage symptoms.
A) Leaves of T. ochroleucum control (top) and after six weeks of stress by 10 μM Cd^{2+} (bottom). The bar represents 2 cm. B) Leaves of *T. caerulescens* of control (top) after four (bottom) and 20 weeks (middle) of growth on 10 μM Cd^{2+}. After four weeks, the plants were under maximal stress (see Fig. 8), while after about 20 weeks they had reached final Cd-acclimation by visual appearance - compare Fig. 3A for acclimation of photosynthesis.

Fluorescence kinetic measurements

Changes in chlorophyll fluorescence kinetic parameters induced by cadmium stress. Unless stated otherwise, all trends discussed in the following were statistically significant as verified by t-tests. The significance level was at least $P < 0.05$, in most cases it was $P < 0.01$; n=number of analysed leaves (for a comparison on the tissue level the leaf is the physiological unit to be counted) was at least 10 (up to 80), from at least seven (up to 19) plants. The numerical values referred to in the following text are taken from measurements in the second series of experiments with *T. caerulescens*, and from the measurements at 40 µmol photons.m^{-2}.s^{-1}. In principle very similar effects were found in the other species, experiments, and measurements. This agreement is illustrated by comparison of the trends for *T. fendleri* and *T. caerulescens* in Fig. 3, panels A vs. B, and the trends for the different irradiances (40 vs. 120 µmol photons.m^{-2}.s^{-1}) in Fig. 3 vs. Supplement-Fig. 1. This shows that cadmium toxicity has the same mechanisms in sensitive and tolerant plants, but the latter are capable to resist better and to recover.

Cadmium-stressed tissues were characterised by a strongly reduced photochemical efficiency. Treatment of *T. caerulescens* with 10 µM Cd^{2+} for 4 weeks diminished the maximal photochemical efficiency of PSII, as measured by F_v/F_m, by up to 70%. Effective quantum yield of photochemical energy conversion in PSII, as defined by $\Phi_{PSII} = (F_m'-F_t')/F_m'$ (Genty et al. 1989) and effective quantum yield of PSII photochemistry (F_v'/F_m') were affected to a similar extent (Fig. 3).

The complete non-photochemical quenching ($q_{CN} = (F_m-F_m')/F_m$) was much less affected and no clear trend was seen in its values plotted as a function of the time of growth on Cd (Fig. 3). In the palisade mesophyll during growth on 10 µM Cd, the q_{CN} in the relaxation period after actinic light was about 50% reduced compared to the control, while no significant effect (t-test, n=73 leaves from 14 plants with 10 cells analysed in each leaf, $P = 0.05$) was found in the spongy mesophyll. In contrast, q_{CN} during the light period was about 30% elevated compared to the control in the spongy mesophyll, but no significant trend (t-test, n=43 leaves from 14 plants with 10 cells measured in each leaf, $P = 0.05$) was found in the palisade mesophyll.

In Cd^{2+} affected plants saturation of photochemistry was attained at lower irradiances than in the controls (Figs. 3, 4, 5, 8). This was well visible in the parameters F_m/F_p (Figs. 4, 6) and $(F_p-F_0)/(F_m-F_0)$ (Fig. 3). Averaged over the whole 10 µM Cd treatment period, $(F_p-F_0)/(F_m-F_0)$ increased from 0.24 ($\pm$ 0.02) to 0.71 ($\pm$ 0.04) of the control in the palisade mesophyll and from 0.53 ($\pm$ 0.02) to 0.82 ($\pm$ 0.03) in the spongy mesophyll. In addition, the rise to F_p after the onset of actinic light was faster in Cd-affected compared to healthy leaves (Figs 5, 8).

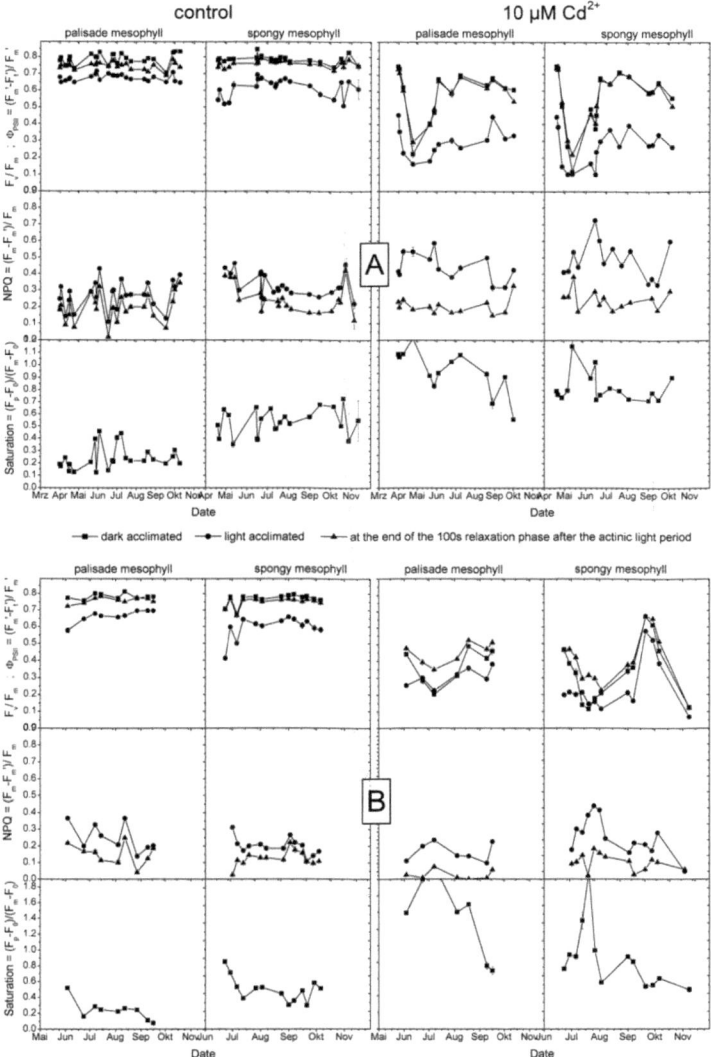

Fig. 3. Changes of fluorescence parameters in young-mature (i.e. just after reaching their final size) leaves during stress and long-term acclimation in response to cadmium treatment. The data shown here are from our second series of experiments (see methods) and were measured with the FKM at an actinic irradiance of 40 μmol photons.m-2.s-1. See Supplement-Fig. 1 for measurements with 120 μmol photons.m-2.s-1 actinic irradiance. Each data point in each figure represents the average and standard error of at least ten individually selected and analysed cells. A) T. caerulescens, B) T. fendleri.

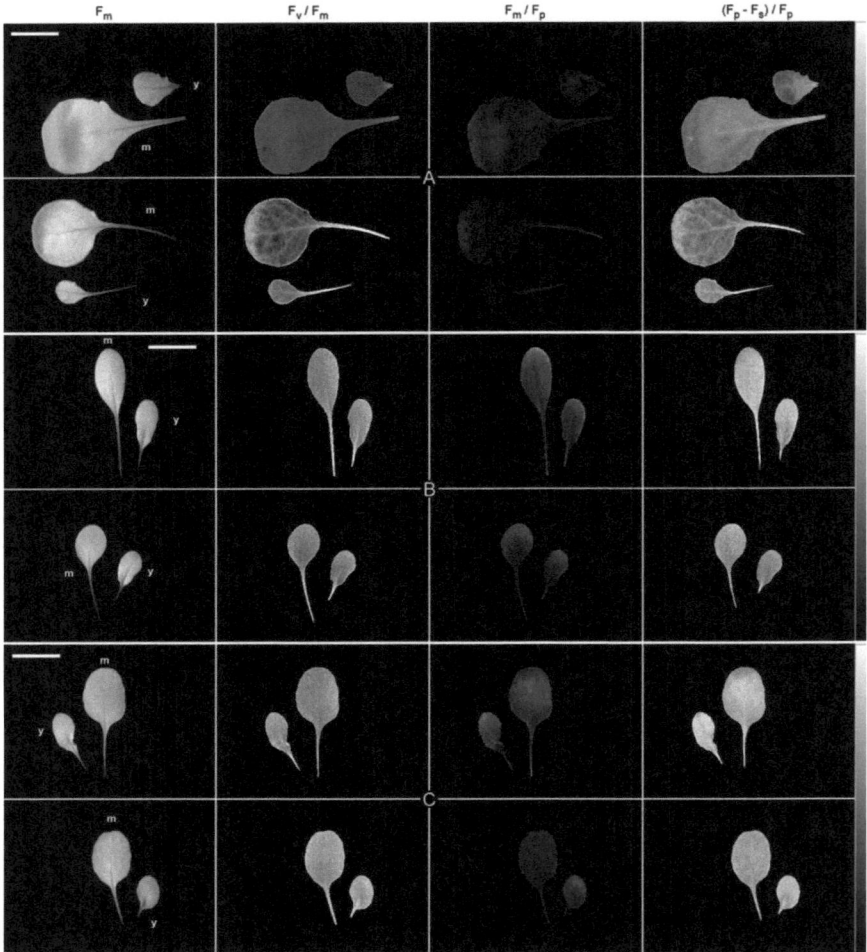

Fig. 4. Macroscopic fluorescence kinetic imaging after three weeks of cadmium treatment: images. The white bar in the upper left image represents 2 cm. The labels in the first panel of each row indicate the age of the individual leaves: y = young, m = mature, at the beginning of senescence. Left column of images: F_m, 2nd column: F_v/F_m; 3rd column: F_m/F_p with an irradiance of about 40 µmol photons.m-2.s-1; right column: $(F_p-F_s)/F_p$ with an irradiance of about 40 µmol photons.m-2.s-1. A) *T. caerulescens*. Top: control; bottom: 50 µM Cd^{2+}. Scale for Fm/Fp: 5 = white, 1 = black. Scale for Fv/Fm: 1 = white, 0.75 = black (the scale was defined only for this small range to make the small changes in Fv/Fm better visible). Scale for (Fp-Fs)/Fp: 0.9 = white, 0 = black B) T. fendleri. Top: control; bottom: 10 µM Cd^{2+}. Scale for Fm/Fp: 3.5 = white, 1 = black. Scale for Fv/Fm: 1 = white, 0.75 = black (the scale was defined only for this small range to show that in contrast to *T. caerulescens* not any heterogeneity of Fv/Fm occurs in this species). C) T. ochroleucum.

Top: control; bottom: 10 µM Cd^{2+}. Scale for Fm/Fp: 3.5 = white, 1 = black. Scale for Fv/Fm: 1 = white, 0.75 = black (the scale was defined only for this small range to show that in contrast to *T. caerulescens* not any heterogeneity of Fv/Fm occurs in this species). Scale for (Fp-Fs)/Fp: 0.9 = white, 0 = black.

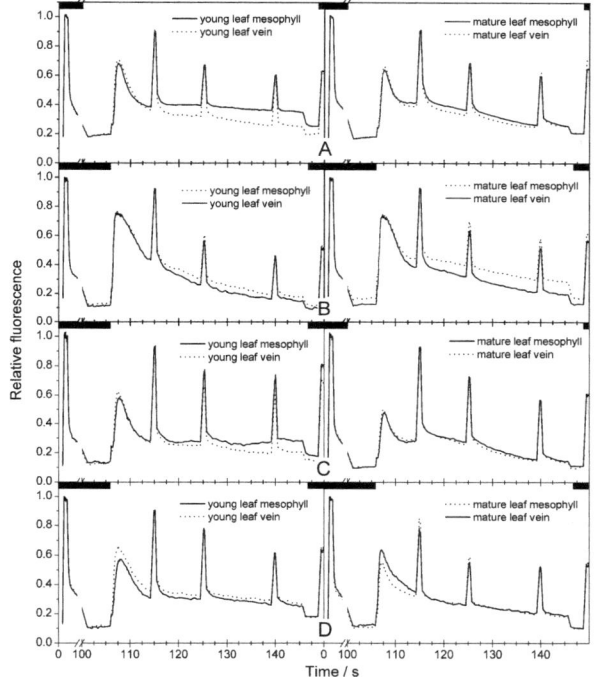

Fig. 5. Macroscopic fluorescence kinetic measurements of *T. caerulescens* and *T. fendleri* stressed with metals for 4 wk. Areas for analysis were chosen in the mesophyll far away from veins (i.e. the most affected area of the leaves) and the central vein (i.e. the least affected area) to show the maximum range of macroscopically measurable variability. Actinic irradiance during measurement: about 40 µmol photons.$m^{-2}.s^{-1}$. The curves have been normalised by their F_m values because among all fluorescence parameters, the dark-adapted F_m best reflects the PSII associated chlorophyll content of the leaf. With the normalisation applied here, changes of F_v/F_m appear as a higher value of F_0. The conventional normalisation by F_0, in contrast, would yield lower values for all types of F_m peaks in strongly affected cells and equal F_0. But independent of the normalisation, of course, are all the ratio parameters. **A)** *T. caerulescens* control, **B)** *T. caerulescens* stressed with 50 µM Cd^{2+}, **C)** *T. fendleri* control, **D)** *T. fendleri* stressed with 10 µM Cd^{2+}. F_v, variable fluorescence; F_m, maximum fluorescence yield of dark-adapted sample; F_0, minimal fluorescence yield of a dark-adapted sample.

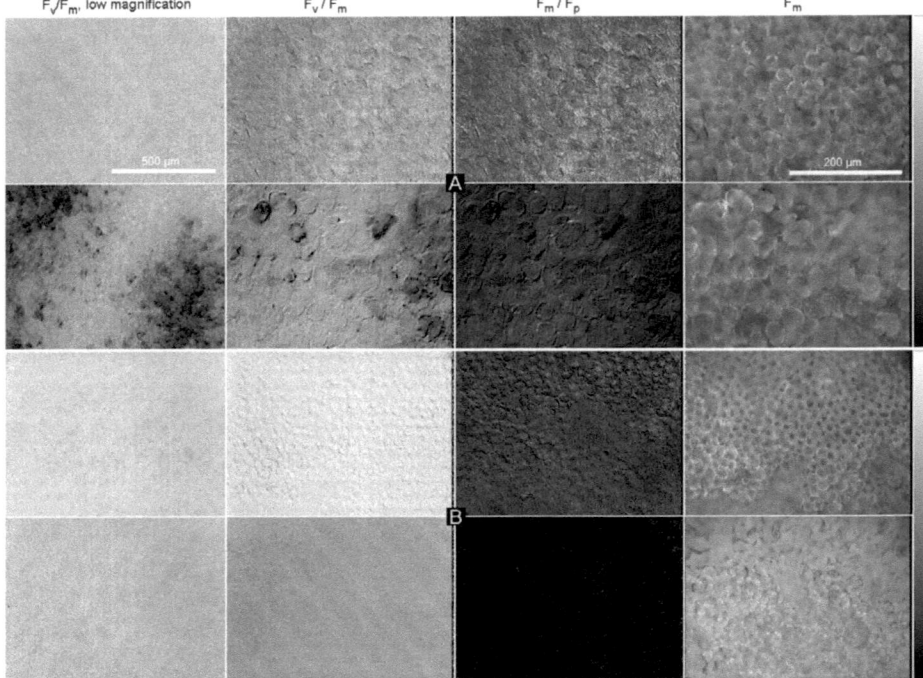

Fig. 6. Fluorescence kinetic microscopy: images.
Left column: F_v/F_m at low magnification, scale: 0.9 = white, 0.3 = black. *Second column*: F_v/F_m, scale: 0.9 = white, 0.3 = black. *Third column*: F_m/F_p *Right column*: F_m. **A)** *T. caerulescens*. The scale bar in the upper left image is valid for all images in that column; the white bar in the upper right image is valid for all other images. *Upper row,* control; *lower row,* treated with 50 μM Cd^{2+} for 5 wk. Scale for F_m/F_p: 5 = white, 1 = black. Actinic irradiance during measurement: about 30 μmol photons.$m^{-2}.s^{-1}$. **B)** *T. fendleri*. *Upper,* control; *lower,* treated with 10 μM Cd^{2+} for 5 wk. Scale for F_m/F_p: 2 = white, 1 = black. Actinic irradiance during measurement: about 200 μmol photons.$m^{-2}.s^{-1}$. F_v, variable fluorescence; F_m, maximum fluorescence yield of dark-adapted sample; F_p, fluorescence yield at the *P* level of the induction curve after the onset of actinic light exposure.

Long-term acclimation to cadmium in *T. caerulescens*. In the beginning of the cadmium treatment, *T. caerulescens* reacted with severe symptoms of Cd toxicity, but after six weeks the photochemistry started to recover (Fig. 3a, Supplement-Fig. 1a). The recovery of most photochemical parameters was rather fast and complete acclimation to 10 μM Cd^{2+} was reached about two months after the beginning of the Cd^{2+} treatment (Fig. 3a, Supplement-Fig. 1a). However, there seems to be some difference in the parameters reflecting predominantly F_v/F_m or

F_v'/F_m' and the Φ_{PSII}s that include the coefficient $\Delta F/F_v'$. With the latter the recovery at 10 μM Cd^{2+} seems to be biphasic: the fast part, parallel to the recovery of F_v/F_m, does not lead to the pre-stress values. From the time point of complete acclimation of F_v/F_m, Φ_{PSII} continues to rise very slowly. The Φ_{PSII} calculated using the florescence response to the first saturating irradiation pulse after the end of the actinic light period assumes a somewhat intermediary value, showing that in the recovered cells a reduced fraction of open PSII reaction centres persists in the first seconds of darkness (supplement-Fig. 1a).

Palisade and spongy mesophyll were affected to a similar extent and followed the same time course of stress and acclimation (compare the left vs. right column of graphs for each treatment in Figs. 3a and 3b). The only observable difference between these tissues was the higher light saturation in the spongy mesophyll, as reflected by the values of $(F_m-F_0)/(F_p-F_0)$. This was to be expected because this tissue should be more shade-adapted than the palisade mesophyll (Fig. 3). Light saturation reached a peak in the initial period of stress, and declined during acclimation.

Nonphotochemical quenching generally did not at all follow the course of events from stress to acclimation that was found for the photochemical quenching parameters (Fig. 3a, Supplement-Fig. 1a). Interestingly, the visible symptoms of damage (see first paragraph of the results) persisted much longer than the inhibition of photochemistry. Similarly, the plants resumed strong growth only after about four months.

In contrast to the 10 μM Cd treatment, 50 μM Cd^{2+} applied at the high nutrient solution flow rate in our second series of experiments (see methods) was lethal to most individuals of *T. caerulescens*. But even in this case a transient recovery was observed before the physiology became so rapidly inhibited that it proved fatal to the plant (Supplement-Fig. 1A, right column).

Time course of inhibition in T. fendleri. In *T. fendleri*, only a transient recovery was observed even at 10 μM Cd^{2+} (Fig. 3b, Supplement-fig. 1b); already this concentration ultimately led to death at the high media flow rate (see methods). In contrast to *T. caerulescens*, in *T. fendleri* NPQ in the light-acclimated state was up to double (at 40 μmol photons.m^{-2}.s^{-1} actinic irradiance, Fig. 3b) compared to the control during the initial period of inhibition.

Macroscopic and microscopic heterogeneity of heavy metal induced stress. Macroscopic imaging of *T. fendleri* and *T. ochroleucum* leaves did not show much heterogeneity in the values of F_v/F_m (Fig. 4). Only in *T. caerulescens* PSII activity was heterogeneous, even in the control plants; mesophyll cells close to veins were more active than those far away from veins. This heterogeneity became more pronounced in Cd^{2+} treated individuals. Remarkably, the average quantum yield (F_v/F_m) of Cd-acclimated *T. caerulescens* did not change much in comparison to the controls, since the decreased activity of the mesophyll cells distant to veins was paralleled by an increased (compared to the control!) activity of cells close to veins (Fig. 4). The difference in fluorescence characteristics of mesophyll and bundle sheath cells (Figs. 4, 5) was, however, much smaller than the differences in their chlorophyll content (Fig. 2). It appears that damaged photosynthetic units quickly bleach out completely, after which they don't contribute to parameters like F_v/F_m. And in bleached leaves the lower F_m per cell may be almost completely compensated, on the macroscopic scale, by deeper penetration of exciting light and fluorescence emission from deeper leaf layers. This is in line with results of non-imaging fluorescence kinetic measurements on copper-stressed plants (Ouzounidou et al. 1997).

Imaging microscopic measurements of fluorescence kinetics showed that in the Cd-sensitive species *T. fendleri* and *T. ochroleucum*, Cd induced damage affected all mesophyll cells approximately equally (Figs. 6b, 8, 9). If any heterogeneity was found, it usually had the form of a shallow gradient from more affected to less affected cells (Fig. 6). In contrast, in the Cd-resistant hyperaccumulator *T. caerulescens*, a few mesophyll cells distributed over the whole leaf surface were strongly inhibited, while the other cells remained much less affected (Figs. 6a, 8, 9). The affected cells sometimes formed clusters. The little affected *T. caerulescens* cells displayed values of the maximal PSII efficiency (F_v/F_m) similar to the cells of an unstressed leaf. Compared to the control, their fluorescence yield was, however, noticeably increased during the first 20 s of actinic irradiance, while their NPQ was distinctly more pronounced (Fig. 8a). The fluorescence kinetics of the inhibited cells were characterised by the features described in the paragraph on changes in fluorescence kinetic parameters induced by cadmium stress, including the increased response to actinic irradiance. Since $F_m/F_p = 1$ equals black in the grey scale, they appeared as dark areas or spots on images of F_m/F_p (Fig. 6).

Qualitative visualisation of Cd-accumulation via a Cd-specific fluorescent dye demonstrated that the spatial pattern of Cd-induced inhibition in the mesophyll was the same as the pattern of Cd-accumulation (Fig. 7). This is in agreement with earlier quantitative data on heterogeneous metal distribution in the mesophyll of hyperaccumulators under metal toxicity stress (Küpper et al. 1999, 2001).

Finally, the spatial heterogeneity of Cd^{2+} induced inhibition of photosynthesis in *T. caerulescens* gradually disappeared when the plants acclimated (Fig. 9).

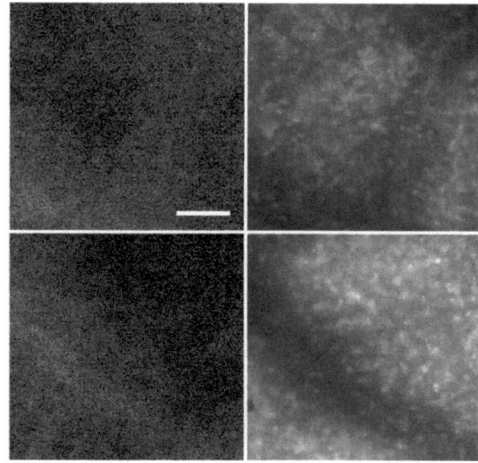

Fig. 7. Two examples of the correlation between heterogeneity of photosynthesis and heterogeneity of Cd^{2+} accumulation in the same mesophyll of a young-mature leaf of *T. caerulescens*. The plant was grown on 10 μM Cd^{2+} for c. 10 wk (i.e. it was in the early acclimation stage). Bar, 200 μm.
Left column: F_v/F_m, measured as for the previous figures. Grey scale from 0 (black) to 0.9 (white).
Right column: qualitative visualisation of cadmium accumulation. This was measured with the cell-permeable dye Rhod-5N, which was infiltrated into the leaf at a concentration of 100 μM before both measurements (F_v/F_m and Cd^{2+}) were done. Grey scale from no Cd (black) to highest Cd^{2+} (white). F_v, variable fluorescence; F_m, maximum fluorescence yield of dark-adapted sample

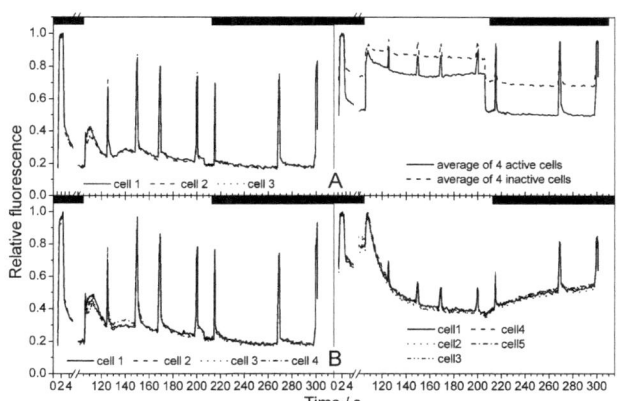

Fig. 8. Fluorescence kinetic microscopy kinetics of objects selected on images like the samples shown in Fig. 6. *Left column:* controls; *right column:* metal stressed plants. The curves have been normalised by their F_m values because among all fluorescence parameters, the dark-adapted F_m best reflects the photosystem II (PSII)-associated chlorophyll content of the leaf. All curves are normalised to $F_m = 1$ for better comparison (see legend to Fig. 5 for explanation). **A)** *T. caerulescens* treated with 50 μM Cd^{2+} for 5 wk. Actinic irradiance during measurement, 120 μmol photons.$m^{-2}.s^{-1}$. **B)** *T. fendleri* treated with 10 μM Cd^{2+} for 5 wk. Actinic irradiance during measurement: 200 μmol photons.$m^{-2}.s^{-1}$.

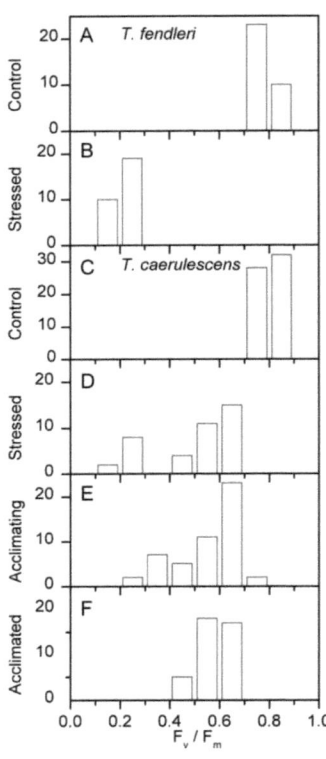

Fig. 9. Histograms of the distributions of maximal photochemical efficiency (F_v/F_m) in *T. fendleri* (A, B) and *T. caerulescens* (C-F), both controls and those treated with 10 μM Cd^{2+}. The cells are a representative subset of all cells visible on the fluorescence kinetic microscope (FKM) records ("films") measured. Stressed, measurements during the week 3-6 of Cd treatment; acclimating, measurements during the week 5-10; acclimated, measurements during the month 4-6 of treatment.

Microscopic in vivo spectroscopy including spectrally resolved fluorescence kinetic measurements. Microscopic absorbance spectra of the mesophyll away from veins, exhibiting the typical symptoms of Cd toxicity, showed a pronounced decrease of pigment content. The OD_{430nm} was typically around 5 compared to an OD_{430nm} of 9.5 for a healthy leaf of the same size (Fig. 10a). No major Cd-induced changes, however, were observed in the shape of the absorbance spectra.

Static single-cell fluorescence emission spectra revealed an about twofold increase in F_0 quantum yield around 680 nm of Cd-stressed compared to unstressed plants; the same was observed also in the imaging records. This applied to both *T. caerulescens* and *T. fendleri* (Fig. 10b) and was probably caused by the decreased re-absorption due to the lowered Chl content in the Cd-stressed leaves. In contrast, the effect of Cd on the fluorescence quantum yield at longer wavelengths (>700nm) was different for both species. In *T. caerulescens*, an increased fluorescence quantum yield was found like for the shorter wavelengths, so that the shape of the emission spectra was about the same for the Cd-stressed leaves and for the controls. But in *T. fendleri* the fluorescence yield above 710 nm decreased in response to Cd-stress, leading to a pronounced change in the shape of the emission spectra (Fig. 10b).

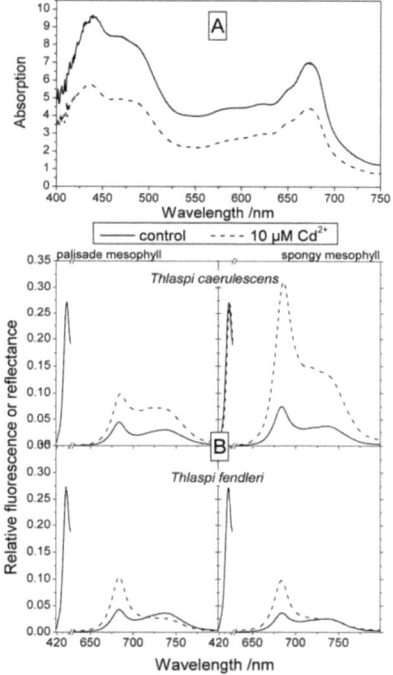

Fig. 10. Static single-cell absorption and fluorescence spectra: control plants and plants grown with 10 µM Cd^{2+} for c. 3 months. (A) Absorption spectra of *T. caerulescens*. (B) F_0 fluorescence spectra. The first peak at about 430 nm represents the crosstalk-peak of exciting light that was used for normalisation.

The mechanism behind these changes was further analysed by the records of spectrally resolved measurements of *in vivo* chlorophyll fluorescence kinetics.

But these measurements yielded a surprise already with healthy plants, which showed a constant value of variable fluorescence quantum yield over the whole spectral range of Chl fluorescence, i.e. 650-800 nm (Fig. 11). This is in sharp contrast to the general view that variable fluorescence should be emitted only by the pigment protein complexes coupled with Photosystem II and that fluorescence emission > 720 nm originates mainly from PS I.

In the Cd stressed plants F_v/F_m was reduced (compared to the control plants) much more in the peaks of Chl fluorescence emission (about 680 nm and 730 nm) than at both shorter (<670 nm) and longer (>750 nm) wavelengths (Fig. 11, left column). This was statistically significant (t-test, n =7 cells in 7 leaves of 7 plants in 3 experiments, P < 0.01) and the effect was much stronger in the palisade (Fig. 11, left column) than in the spongy mesophyll (Fig. 11, right column). As also shown in Fig. 11, with spectrally resolved values of $\Phi_{PSII} = (F_m' - F_t')/F_m'$ this surprising behaviour was even more pronounced (t-test, n = 7 cells in 7 leaves of 7 plants in 3 experiments, P < 0.01), in particular for the Cd-acclimated plants. A completely different spectral distribution of Cd-induced inhibition was found for NPQ. In both palisade and spongy mesophyll, the Cd-induced changes of this parameter were strongest around 680 nm, where NPQ was much increased, while at > 700nm Cd had less effect on NPQ. While the wavelength-dependency of the Cd-induced NPQ increase was more pronounced in the palisade mesophyll, the average NPQ increase was stronger in the spongy mesophyll (bottom row of Fig. 11). This enhancement and spectral heterogeneity of NPQ was found in both stressed and acclimated plants.

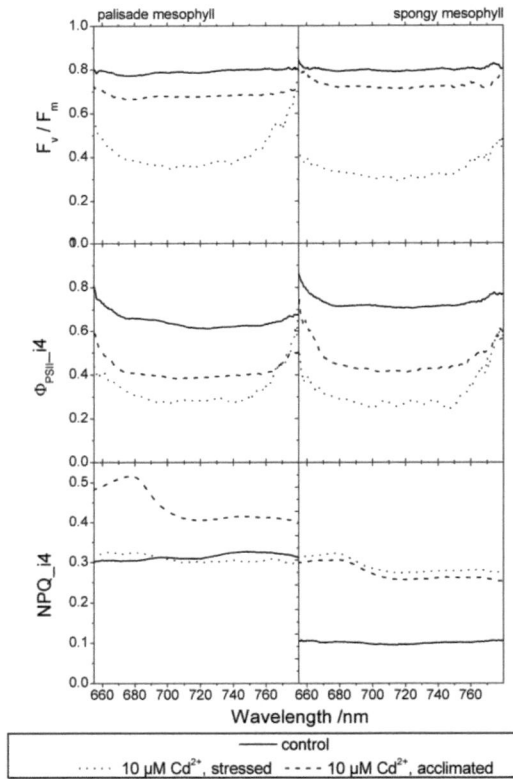

Fig. 11. Spectrally resolved single-cell Chl fluorescence kinetics of *T. caerulescens*: control plants and plants grown with 10 µM Cd^{2+} for c. 2 months (stressed state) and c. 4 months (early acclimated state). Measurements were done with an actinic irradiance of c. 40 µmol photons.$m^{-2}.s^{-1}$. The graphs of the parameters are averages of usually four to 10 individual measurements and are shown with 11-point gliding average smoothing. F_v, variable fluorescence; F_m, maximum fluorescence yield of dark-adapted sample; NPQ, nonphotochemical quenching; Φ, effective quantum yield of photochemical energy conversion in actinic light.

DISCUSSION

Unfortunately, not many papers are available reporting on fluorescence characteristics in plants supplied with heavy metals in ecologically significant concentrations including the sublethal and physiologically tolerable range. But the effects induced by such "natural" applications often markedly differ from those observed in the most popular measurements using very high concentrations or the *in vitro* heavy metal applications to isolated thylakoids or chloroplasts (for a recent review see e.g. Küpper & Kroneck; 2005). Working with concentrations relevant for the environment of hyperaccumulators, this study yielded new results mainly in two areas: knowledge of the mechanisms of Cd-induced inhibition of photosynthesis and physiological acclimation to Cd-induced stress.

In hyperaccumulators a strong barrier evidently separates mesophyll cells from epidermal ones, as the latter serve as the principal dumping site for heavy metals in extremely high concentrations (see introduction) while the former are active in photosynthesis. But the mesophyll of hyperaccumulators still contains much more metal than that of normal plants (Küpper et al. 1999, 2000a, 2001). And the current study revealed that shortly after the onset of heavy metal stress, damage to the mesophyll occurs. This stress may occur in nature whenever the roots of hyperaccumulators become long enough to penetrate through the leached-out top few centimetres of soil into the heavy metal rich soil below (see introduction). Our results have shown that the resistance mechanisms of hyperaccumulators are able to cope well even with this situation.

Mechanisms of Cd-induced inhibition of photosynthesis during stress and acclimation

At first glance, our measurements of cadmium-induced effects on acclimated leaves of *T. caerulescens* qualitatively agree with results of Krupa *et al.* (1993) obtained with *Phaseolus vulgaris*. The time-course of inhibition and quantitative aspects of the data are, however, different and lead us also to different conclusions. Seemingly in agreement with Krupa *et al.* (1993, 1999) our records of acclimated plants show that the maximum quantum yield of PSII photochemistry, F_v/F_m, was not much affected. This is different when looking at the time course of Cd-induced inhibition (Fig. 3). The high F_v/F_m in the acclimated state was in contrast with a dramatic decline of this parameter during the initial period of stress, and for lethal Cd treatments before the death of the plant. Thus Cd does strongly impair the primary events in PSII RC. In line with Krupa *et al.* (1993), photochemical quenching in actinic light (Φ_{PSII}, F_v'/F_m') noticeably decreased in the Cd^{2+} treated plants. In addition, in Cd-stressed plants we observed an increased saturation, as measured by the increased amplitude of F_p relative to F_m. F_p/F_m and $(F_p-F_0)/(F_m-F_0)$ are mostly dependent on the ratio of functional antenna molecules to functional reaction centres and electron transport chains. Under constant actinic irradiance for measuring F_p, a large antenna capturing photons and delivering them to its reaction centre will cause more of the "electron traffic jam" that leads to F_p than a small antenna. Therefore, these changes indicate a reduced ratio of functional PSII RCs (or functional PSII - PS I electron transport chains) to LHCII in a system with high antenna connectivity, which is in line with the reduced values of F_v/F_m and indicates damage in the PSII RCs. Such effect has been suggested earlier for the *"sun reaction"* type of heavy metal-induced inhibition of photosynthesis (Küpper et al. 1996, 1998, 2002). In contrast to Krupa *et al.* (1993), in our experiments the values of NPQ were less affected by Cd^{2+} treatment than the values of photochemical quenching. Therefore, we cannot follow the explanation proposed by Krupa *et al.* (1993) that the main cause of Cd-induced inhibition is the reduction in the rate of the Calvin-Benson cycle, followed by down-regulation of PSII photochemistry. The results of the current study strongly suggest that the light reactions were more severely inhibited by Cd than the Calvin-Benson Cycle. A simple explanation of the effects on fluorescence that we observed (long-term elevated NPQ, slower recovery of Φ_{PSII} compared to Fv/F_m) would be that besides a strong inhibition of the PSII RCs in the peak of Cd stress, the passage of electrons through the cyt b_6f complex is slowed down for a much longer period. The reason for this difference, however, remains a topic for future studies. The differences to the study by Krupa *et al.* (1993) may be caused by different experimental conditions or different plant genera (*Phaseolus* vs. *Thlaspi*).

Cd-induced chlorosis was much more pronounced on young leaves that emerged after the onset of acute Cd stress than on old leaves that were present already before. This could indicate that young leaves take up more Cd, or that they are more vulnerable to Cd toxicity. The latter is in line with the interpretation of Krupa *et al.* (1993, 1999) that Cd inhibits Chl biosynthesis, but could alternatively also be caused by a better accessibility of Chl to Mg-substitution by Cd^{2+} in young leaves. Cd-Chl is highly unstable and would decay (bleach) shortly after formation (Küpper et al. 1996).

The spectrally resolved single-cell fluorescence kinetic measurements (Figs. 11) yielded a novel information on variable fluorescence in healthy *Thlaspi* leaves, namely that the yield of F_v remains constant even beyond ~710 nm . The literature is dominated by the idea that Chl fluorescence in this spectral region is mainly emitted by PSI that does not display variable fluorescence characteristic of PSII. Lower variable fluorescence for the spectral region above 710 nm compared to the spectral region around 680 nm has been recorded among others by Ruban & Horton (1994) and Franck *et al.* (2002, 2005). The difference to our results might be due to differences in plant material or methods applied. Nobody as yet applied spectrally resolved kinetic fluorescence analysis to one cell. It should be also noted that the earlier studies did not derive variable fluorescence from complete Kautsky kinetics including the response to saturating flashes.

Finally, at least in the study of Franck *et al.* (2005), the shape of the spectrum, in which F_m/F_0 is low wherever fluorescence is low, could indicate a problem with insufficient background (dark current, crosstalk) subtraction.

The application of spectrally resolved single-cell fluorescence kinetic measurements on the analysis of Cd-stressed plants yielded further insights into the mechanism of Cd-induced damage. The spectral differences in the reduction of photochemical activities (F_v/F_m, Φ_{PSII}) in Cd affected cells could lead to the assumption that Cd toxicity produced non-functional antenna complexes emitting non-variable ("dead") fluorescence that elevates F_0 and thereby reduces F_v/F_m. Such an inhibition of the antenna is known from heavy metal (mainly Cu and Zn) induced damage in low irradiance with a dark phase and was termed *"shade reaction"* (Küpper et al. 1996, 1998, 2002). In the present case, however, this explanation is improbable, because the "inhibition band" is much broader (680 - 740 nm) than the emission maximum of LHCII (around 680 nm). Therefore, the reduced F_v/F_m is most likely caused by inhibition of the function of the PSII RC, which affects the fluorescence quantum yield of all Chls in the antenna coupled to it. Cd-induced inhibition of the PSII RC has been proposed by many other authors (for reviews see e.g. Prasad & Hagemeyer, 1999; Küpper & Kroneck, 2005). The significantly lower Cd-induced inhibition of the margins of the Φ_{PSII} spectrum compared to the region 680-750 nm could be due to effects of intracellular self-absorption. In these margins, *in vivo* Chl absorption is much lower than in the centre, so that the fluorescence of chloroplasts deeper in the cell will contribute more to a fluorescence signal measured with a non-confocal system as ours. These deeper chloroplasts receive less actinic light due to overlaying chloroplasts, so that their Φ_{PSII} is higher than that of chloroplasts closer to the leaf surface. The observation of that reduced sensitivity of the spectral margins also concerns F_v/F_m (a parameter not influenced by actinic irradiance), however, indicates that in addition chloroplasts exposed to more light are inhibited by Cd more strongly than shaded ones.

The pronounced peak at 680 nm in the increasing NPQ in Cd-stressed plants in comparison to the broad spectral band of inhibition of F_v/F_m and Φ_{PSII} is particularly interesting. First, the different spectral distribution of the parameters shows that the Cd-induced inhibition of photochemistry and the increase of NPQ have different mechanistic reasons. They clearly influence chlorophylls emitting or quenching fluorescence at different wavelengths, i.e. Chls in different pigment-proteins or in different binding environments in the same protein. Further, the absence of the 680 nm peak in NPQ of control plants shows that the mechanism of NPQ itself is altered by Cd stress. An antenna complex fluorescing with a peak at 680 nm seems to contribute to NPQ more strongly (or only) under Cd-induced stress. This antenna could be the LHCII, which would normally deliver excitons to PSII. After the destruction of some PSII reaction centres (see above), the delivery of excess excitons to the remaining PSII RCs could be prevented by an enhanced NPQ specifically quenching LHCII excitons. Such a reduced fluorescence quantum yield of LHCII could result from a state transition, after which LHCII would deliver its excitons to PS I without emitting fluorescence. LHCII-mediated NPQ has been suggested already by Ruban & Horton (1994) based on spectroscopic data. However, they only analysed healthy *Guzmania* plants and found two NPQ peaks (683 nm and 698 nm). In contrast, our healthy *Thlaspi* did not show these peaks at all. And also Cd-stressed *Thlaspi* did not exhibit two NPQ peaks, but a single NPQ peak at about 680 nm in agreement with fluorescence emission spectra measured on isolated LHCII (Hemelrijk et al. 1992). Since fluorimeters usually measure Chl fluorescence in the 680 nm peak, much of the additional Cd-induced NPQ measurable in Cd-stressed plants (see above and Krupa et al. 1993) may be due to this NPQ mechanism. This enhancement of NPQ, however, remained even after the plants acclimated to the Cd-stress and resumed strong growth. This shows that the damage it reflects is not the main damage mechanism during the period of stress.

Spatial heterogeneity of photosynthesis during acclimation to Cd-induced stress

Macroscopic imaging indicated that mesophyll cells close to veins were more cadmium resistant than the mesophyll in between. So either the metal reached these cells only after being released from the veins at their tips and having passed through the mesophyll, or under stress the plants actively try to keep the phloem of the veins functioning and rather sacrifice some of the mesophyll.

Microscopic imaging revealed that a more important photosynthetic heterogeneity exists on the single-cell level. In stressed *T. caerulescens*, only few mesophyll cells were damaged while most others stayed healthy, and this heterogeneity disappeared when the plants acclimated to Cd. The heterogeneity of photosynthetic activity in *T. caerulescens* directly correlated with the heterogeneity in the accumulation of labile bound Cd^{2+} in the same mesophyll cells. Our measurements of Cd-accumulation based on a Cd-specific fluorescence dye may be only qualitative, but they can be regarded as reliable because they are backed by quantitative results from earlier studies on compartmentation of Cd^{2+} in *Arabidopsis halleri* (Küpper et al. 2000a) and Ni^{2+} in *Thlaspi goesingense* and *Alyssum lesbiacum* (Küpper et al. 2001). Those measurements showed that under stress conditions an increased accumulation of heavy metal (Cd^{2+} or Ni^{2+}) occurs in some of the mesophyll cells. The same cells also accumulated elevated (compared to neighbouring normal mesophyll cells) concentrations of Mg^{2+}. This was interpreted as a defence against the substitution of Mg^{2+} in Chl by heavy metals. A recent semi-microscopic study on metal accumulation (Cosio et al. 2005) found spots of Cd accumulation on leaves of Cd-stressed *T. caerulescens*.

The Cd-sensitive species *T. fendleri* and *T. ochroleucum* displayed less heterogeneity in photosynthesis on the cellular level. The occurring damage affected most cells of the leaves and usually only a shallow gradient from more to less affected cells was found.

All this indicates that the heterogeneity between individual cells is an emergency defence mechanism of *T. caerulescens* against Cd-induced stress. Under moderate metal load, the sequestration of metals in the vacuoles, mainly of epidermal cells, is sufficient. When the transport into these cells is temporarily overloaded by inhibitory metal concentrations, damage occurs in the mesophyll. In hyperaccumulators, under such conditions a few mesophyll cells are sacrificed and used as metal dump. This results in the heterogeneity of metal accumulation reflected by the heterogeneity of photosynthetic characteristics. If a plant thus survives, this emergency defence mechanism is again replaced (probably involving an enhanced expression of transporters) by the sequestration of metals into epidermal vacuoles, which abolishes the metal toxicity almost completely. The regulatory mechanism responsible for the selective inhibition that is actively kept at non-lethal level is an important task for current further investigations.

ACKNOWLEDGEMENTS

We would like to thank Sven Seibert for help with growth of plants and recording of data in the second series of experiments, Anna Ruprechtová and Marie Šimková for help with plant growth in the first series of experiments, and Todor Macho for precision mechanical work on the FKM. We are grateful for a generous donation of various research instruments from the Degussa AG (formerly Hüls AG, Marl). H. Küpper was supported by a fellowship from the Studienstiftung des Deutschen Volkes (Bonn), Aravind Parameswaran was supported by a fellowship of the German Academic Exchange Service (DAAD). The research was further efficiently supported by the grants VS96085 and ME138 from the Ministry of Education of the Czech Republic, a grant from the Stiftung Umwelt und Wohnen and grant 661278 from the Fonds der Chemischen Industrie (FCI).

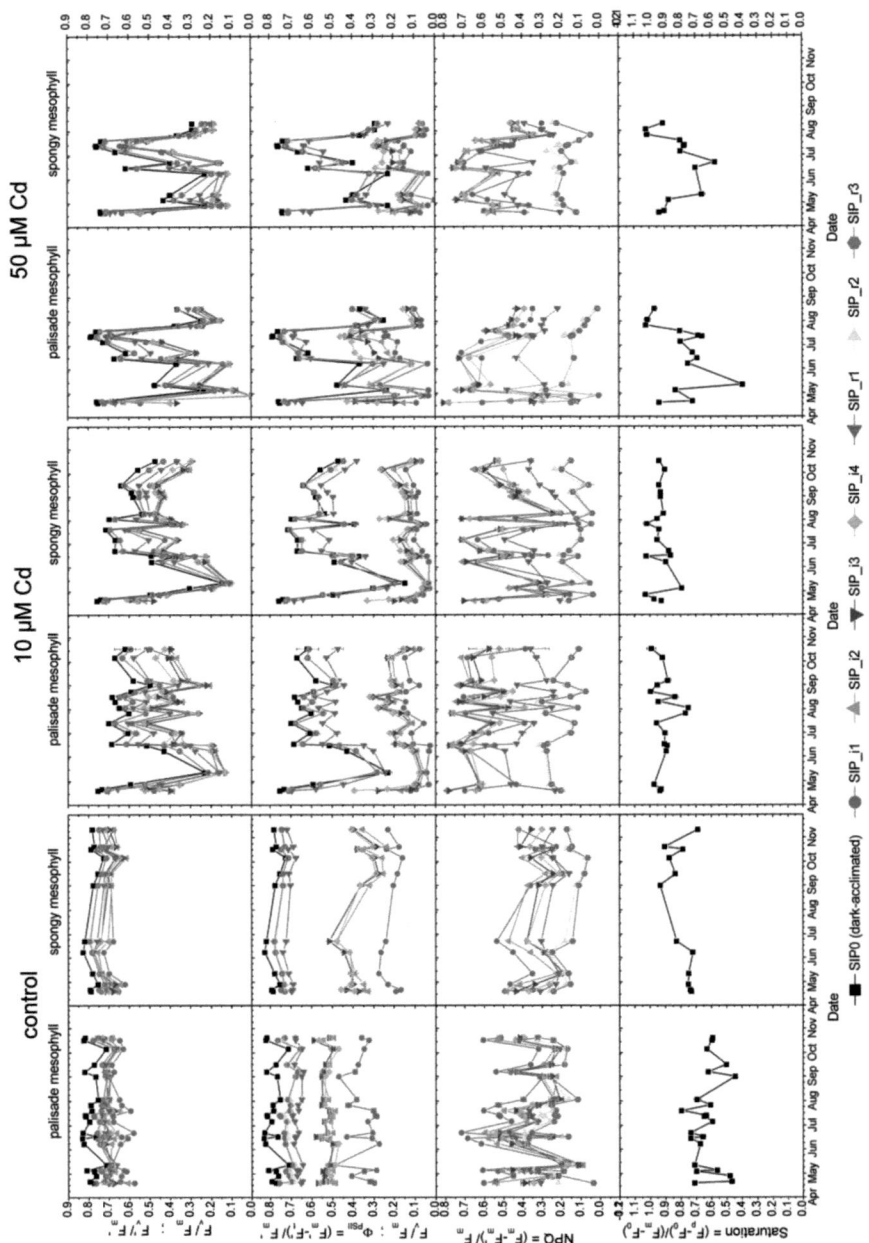

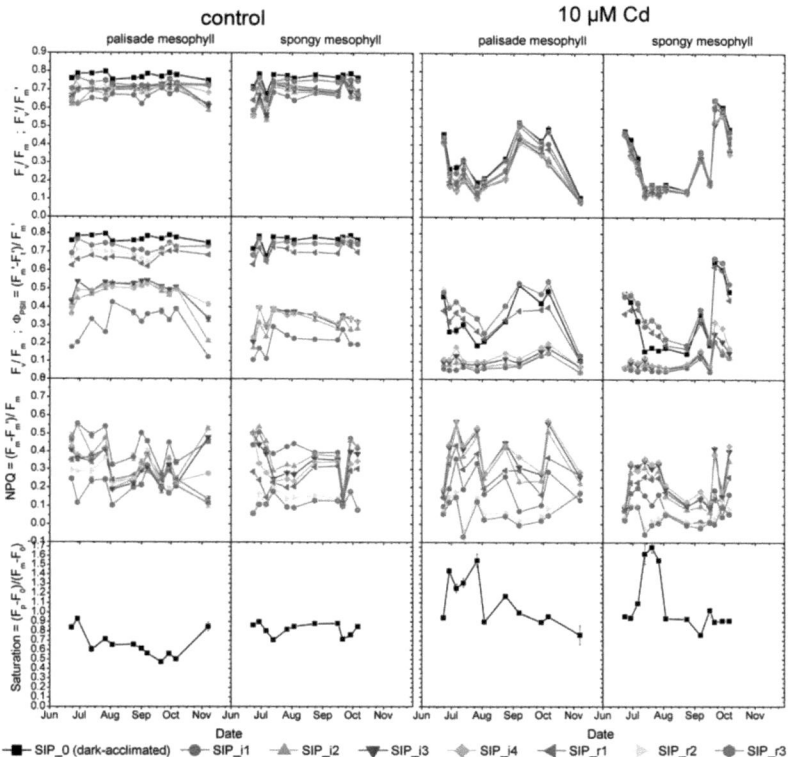

Supplement-Fig. 1. Changes of fluorescence parameters in young-mature (i.e. just after reaching their final size) leaves during stress and long-term acclimation in response to cadmium treatment, measured with the FKM at an actinic irradiance of about 120 µmol photons.m^{-2}.s^{-1}. "SIP" means "saturating irradiation pulse, they are numbered in their sequence of measurement. "i" stands for the period of actinic irradiation, "r" for the period of dark relaxation. Each data point in each figure represents the average and standard error of at least ten individually selected&analysed cells. The data are from the second series of experiments (see methods) A) *T. caerulescens*, B) *T. fendleri*.

2.2. Cadmium uptake and sequestration kinetics in individual leaf cell protoplasts of the Cd/Zn hyperaccumulator *Thlaspi caerulescens*

Barbara Leitenmaier [a] and Hendrik Küpper [a,b, 1]

[a] Universität Konstanz, Fachbereich Biologie, D-78457 Konstanz, Germany

[b] Faculty of Biological Sciences and Institute of Physical Biology, University of South Bohemia, Branišovská 31, CZ-370 05 České Budejovice, Czech Republic

Corresponding author: Hendrik Küpper, Fachbereich Biologie, Universität Konstanz, Universitätsstrasse 10, D-78464 Konstanz, tel. ++49-7531-884112, fax. ++49-7531-884533, email: hendrik.kuepper@uni-konstanz.de

published in Plant, Cell & Environment (2011) 34: 208-219

ABSTRACT

Hyperaccumulators store most of the accumulated metal in the vacuoles of large leaf epidermal cells (storage cells). For investigating cadmium uptake, we incubated a protoplast mixture obtained by digestion of leaves of *Thlaspi caerulescens* (Ganges ecotype) with a Cd-specific fluorescent dye. A fluorescence kinetic microscope was used for selectively measuring Cd-uptake and photosynthesis in different cell types, so that a physical separation of cell types was not necessary. Few minutes after its addition, cadmium accumulated in the cytoplasm before its transport into the vacuole. This demonstrated that vacuolar sequestration is the rate limiting step in cadmium uptake into protoplasts of all leaf cell types. During accumulation in the cytoplasm, Cd-rich vesicle-like structures were observed. Furthermore, Cd uptake rates into epidermal storage cells were higher than into standard sized epidermal cells and mesophyll cells. This shows that the preferential heavy metal accumulation in large epidermal storage cells, previously observed for several metals in intact leaves of various hyperaccumulator species, is due to differences in active metal transport and not differences in passive mechanisms like transpiration stream transport or cell wall adhesion. Combining the current results with previous studies, it seems likely that the transport steps over the plasma and tonoplast membranes of leaf epidermal storage cells are driving forces behind the hyperaccumulation phenotype.

KEYWORDS

Cadmium uptake, Fluorescence Kinetic Microscopy, fluorescent dye, metal sequestration, storage cells, *Thlaspi caerulescens*

INTRODUCTION

Cadmium has been found to be a micronutrient for an ecotype of *Thalassiosira weissflogii*, a marine alga (Lane and Morel, 2000) and many other heavy metals such as copper, nickel and zinc are well-known for a long time already as essential trace elements for plants. Cadmium can occur in very high concentrations that are detrimental, in many cases even lethal to most plant species, as a result of various human activities (Buchauer, 1973; Fergusson et al. 1980; McBride et al. 1997; VanGeen et al. 1997). Above the threshold leading to growth inhibition by heavy metals, a variety of toxic effects have been observed in cyanobacteria as well as in plants, as described in a comprehensive review on this subject (Prasad and Hagemeyer 1999) and in a more recent one (Küpper and Kroneck 2005). Some plants, called hyperaccumulators, actively take up large amounts of potentially toxic metals and store them in their above-ground tissues (first described by Risse in the article of Sachs, 1865, term "hyperaccumulator" introduced by Brooks, 1977). While part of the hyperaccumulation phenotype is due to enhanced root to shoot translocation as shown by electrophysiology (Lasat et al. 1996, 1998) and studies with radioactive cadmium (Zhao et al. 2006), most of the metal is stored in the above-ground parts. Hyperaccumulators have to store the taken up metal in a way that it does not harm important enzymes and especially not photosynthesis, therefore it is important to keep the metal concentration in the cytoplasm of mesophyll cells as low as possible. It makes sense for hyperaccumulating plants to store metal in the vacuoles because this organelle only contains enzymes like phosphatases, lipases and proteinases (Wink, 1993) that were never identified as targets of heavy metal toxicity. Additionally it has been shown that high amounts of metals are stored specifically in the vacuoles of large epidermal cells (Küpper et al, 1999, 2001; Frey et al. 2000), where no chloroplasts are located and therefore photosynthesis cannot be inhibited. These cells were furthermore found to display a strongly elevated expression of the metal transporters MTP1 and ZNT5 (Küpper and Kochian, 2009). Also many previous studies have shown a strongly elevated expression of metal transport genes in hyperaccumulators compared to non-accumulators (first found by Pence et al. 2000; reviewed by Verbruggen; Hermans & Schat 2009). But for most of these genes the cellular expression pattern and its metal-dependent regulation remains unknown. And in some cases an investigation of the cellular expression pattern showed that the elevated expression is not a cause, but rather a consequence of the hyperaccumulation phenotype (Küpper et al. 2007b; Küpper and Kochian, 2009). For long-term storage in the vacuoles, hyperaccumulated metals are bound only to weak ligands like organic acids (Küpper et al. 2004, 2009a). The sequestration into vacuoles is a transport of metal against the concentration gradient and therefore needs an active transport system (Salt & Wagner, 1993). Until now, only a few transporters for vacuolar sequestration of zinc (and possibly cadmium) and nickel have been investigated and could be characterised (Van der Zaal et al. 1999; Elbaz et al. 2006;

Desbrosses-Fonrouge et al. 2005; Haydon and Cobbett, 2007; Morel et al. 2009). It is difficult to study the uptake of different substances into plant cells, because (in contrast to cells from bacteria or animals) they are, surrounded by a cell wall that consists mainly of cellulose. Fluorescence microscopy uses specific dyes to make substances visible, but in plants those dyes often bind to cell walls. Additionally, the emission of those dyes in most cases overlaps with autofluorescence. Finally, the fluorescence response of the dyes to increasing metal concentrations is not at all linear, but displays a strongly sigmoidal shape. All these problems make the use of normal plant cells for detecting the transport of a target substance into the cell very difficult and usually not quantifiable.

To overcome these difficulties, in the current study we applied the following strategies. First, we have chosen a dye that displays high selectivity to Cd (Soibinet et al. 2008) with a strong emission band in a spectral range where plant cells exhibit only very little autofluorescence. Second, we worked on cells without cell wall, i.e. protoplasts. Protoplasts can be isolated by digestion of the cell wall, so that fluorescent dyes can not bind to the cell wall any more. This procedure makes it possible to get insights into the transport of substances over the cell membrane into the cytoplasm, which was shown already for the transport of cadmium into cells from wheat seedlings (Lindberg, Landberg & Greger 2004), and from the cytoplasm into the vacuole. It is important to investigate differences in cadmium transport into mesophyll cells compared to epidermal cells, and furthermore to measure it in epidermal cells of different sizes, as it is known that large epidermal cells are used for the storage of high amounts of metal (Küpper et al.1999, Küpper et al. 2001, Frey et al. 2000). In several cases the attempt to obtain epidermal protoplasts and/or the cadmium detection in epidermal protoplasts of *T. caerulescens* failed (Cosio et al. 2005, Marques et al. 2004, Ma et al. 2005). These failures were, at least in part, due to problems with the gradient centrifugation that was used for physically separating cells from the mesophyll and the epidermis (Cosio et al. 2005). This separation is not necessary when the fluorescence kinetic microscope (FKM) is used where individual cells can be measured in a mixture of cell types (Küpper et al. 2000, 2007a). Furthermore, in this system the cells are kept vital by delivery of fresh medium to the sample during the measurement and by measurement with extremely low irradiances (Küpper et al. 2000, 2007a).

In this study, we investigated Cd-uptake into individual protoplasts in a significant number of replicate protoplast preparations, so that we could perform statistics of uptake into different cell types. In this way, we were able to solve two key questions of Cd accumulation in leaves, which are the main metal storage sites of hyperaccumulator plants: 1. Is the preferential heavy metal accumulation in large epidermal storage cells compared to other cell types due to differences in active metal transport or due to differences in passive mechanisms like the transpiration stream transport or cell wall adhesion? For solving this question, we compared the metal accumulation pattern previously observed for several metals in intact leaves of various hyperaccumulator species (incl. *Thlaspi caerulescens*) with the rates of active metal uptake in cells where the cell wall was removed, i.e. protoplasts. 2. Which is the time limiting step in metal sequestration into large epidermal cells? To answer this question, we supplied protoplasts from plants grown without cadmium with cadmium during the uptake experiments and measured the kinetics of Cd accumulation in different subcellular compartments.

MATERIAL AND METHODS

Plant material and growth conditions

Seeds of *Thlaspi caerulescens* J.&C. Presl (Ganges population) were germinated and plants were grown as described in Küpper et al. (2007a), but with only 0.1 µM Zn in the nutrient solution to diminish problems of detecting transport of Cd on a high background of zinc.

Preparation of protoplasts

Protoplasts were prepared based on a method developed by Coleman, Randall & Blake-Kalff (1997) and modified for *T. caerulescens* as described, in principle, by Ferimazova et al. (2002). Leaves of *T. caerulescens* were cut freshly and first wiped off with 70% (v/v) ethanol. After that, the lower epidermis was removed using a clean razor blade to set a cutting near the leafstalk, and watchmakers forceps were used to carefully strip off the lower epidermis (which was discarded as most storage cells in the stripped-off epidermis break during the stripping). Then the stripped leaf was put into a Petri dish filled with 2.5 ml of sterile filtrated isolation medium (IM) containing 1.0 M D-Sorbitol and 10 mM MES. The pH was adjusted to 5.6 using BTP (Bis-tris-propane). D-Sorbitol and MES were purchased from Merck, Darmstadt (www.merck.de), BTP was purchased from Sigma, Steinheim (www.sigma-aldrich.com). Finally, 2.5 ml of a digestion medium (DM), containing IM plus 0.7% cellulysin, 0.4% macerase, 0.002% pectolyase and 1% BSA were added. The sample was incubated for 16 h at a temperature of ~23°C. To prevent them from damaging the protoplasts, the cuticle and leaf-veins were removed once the attached cell wall had been digested.

Sample preparation

For measuring the uptake of cadmium into cells, we incubated the protoplasts with the fluorescent dye Rhod5N AM (Invitrogen, department Molecular Probes, Eugene, Oregon, USA, www.invitrogen.com). Rhod5N belongs to the acetoxymethyl (AM) ester derivatives of fluorescence indicators; the modification of carboxylic acids with AM ester groups leads to uncharged molecules, which can permeate cell membranes easily. When the molecule enters the cell, these lipophilic blocking groups are cleaved by intracellular esterases and the molecule gets a charge. The charged form can only very slowly leak out of the cell. The hydrolysis of esterified groups is also important for binding the target ion. Rhod5N AM was originally developed for visualising calcium, but its binding and fluorescence response is 2-3 orders of magnitude higher for Cd^{2+} than for Ca^{2+}. This fact was verified in our work to make sure that the detected signal had its origin in the cadmium uptake into the cells and not in the normal calcium content, which is present in all living cells (Fig. 1). Cells were incubated in IM with the addition of 100 µM Rhod-5N AM (dissolved in DMSO to a concentration of 10 mM, then IM was added to reach the final concentration) for 1 h at room temperature in the dark on a shaker. After incubation, the cells were washed three times with IM. In each washing step, cells were centrifuged at 10xg for 1 min, the supernatant was replaced by fresh IM. After washing, the sample was put onto the glass of a measuring chamber (Küpper et al. 2000, 2007a) and carefully mixed with 0.1625% of SeakemGold Agarose (purchased from Cambrex, Rockland, ME USA, www.cambrex.com) in IM to avoid any movements of cells during the measurements. Then the mixture of cells and agarose was covered by cellophane and a silicone o-ring was placed around the cellophane to stretch it and keep it in place. Prior to use, the cellophane was boiled in distilled water for removal of metal contaminations, and then soaked with IM for at least one hour to minimise shrinking/swelling during measurement.

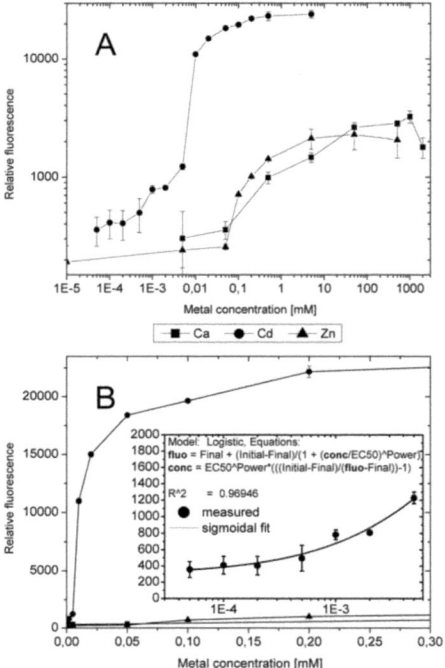

Fig. 1. Calibration of Rhod5N-AM: fluorescence signals obtained for cadmium, zinc and calcium in different concentrations under the same measuring conditions (irradiance, optical system, detector settings) as in the *in vivo* measurements, incubated in isolation medium with the potassium salt of the dye.

(A) Full range of the calibrations, showing the different maximal fluorescence of the three metal complexes of Rhod5N.

(B) Physiological range of metal concentrations and sigmoidal fit of the curve in the micromolar range relevant for short-term (range of hours) metal uptake.

Microscopic measurements

For all measurements in this study, the new version of the "Fluorescence Kinetic Microscope" (FKM) was used, the features of the instrument are explained in detail in Küpper et al. (2007a). For the current work, the main advantage over traditional fluorescence microscopes was the optimised light control and detection system, minimising the photodamage to the cells while yielding quantitative data.

Measurements were, in principle, carried out according to the method used for leaves as described in Küpper et al. (2007a). But in this study the cells where fixed in the measuring chamber using agarose. In contrast to leaves, which can be measured in the presence of air saturated with water, protoplasts have to be kept in IM to keep them vital as long as possible. For this purpose, the measuring chamber as described by Küpper et al. (2000, 2007a) was used, and fresh medium was pumped through the chamber during the measurements as described in Küpper et al. (2008) for studies on cyanobacteria.

Cadmium-uptake-kinetics

Directly before the measurement of Cd-uptake, 10 µM $CdSO_4$ was added to the medium, then the solution was mixed carefully and immediately the measurement was started using a 40x magnifying objective. Rhod5N with bound Cd^{2+} was detected with a special set of filters (from AHF, Germany, www.ahf.de): excitation at a wavelength of 530-550 nm (AHF HQ 545/30), dichroic mirror at 565 nm (AHF 565 DCLP), and emission at a wavelength of 578 - 633 nm (AHF D 605/55).

Two different measuring protocols were used. The first had a total length of 1000 s. In this protocol, after two seconds a brightfield image is taken, then the filter for detection was changed. After 11 s, the signal of Rhod5N AM with bound cadmium was detected for 0.5 s with actinic light (1700 µE) combined with superlight (3900 µE). The lights were switched on again for further Cd measurements every 10 s. After 100 seconds, until the end of the kinetic measurement, the lights were switched on only every 100 s for 0.5 s. For very long measurements, the second protocol with a total length of 3612 s was used. The only difference to the protocol above were the time intervals; only every 100 s the lights were switched on from the beginning. This protocol was only used in special cases, when an object was very stable in its position and was expected not to leave the focus.

Photosynthesis measurements

Photosynthesis measurements were used to obtain information about the vitality of the mesophyll cells. The parameter $F_v/F_m = (F_m-F_0)/F_m$ was measured for assessing the maximal dark-adapted quantum yield of photosystem II. $\Phi_{PSII} = (F_m'-F_t')/F_m'$ was measured to estimate eletron flow through PS II in the light acclimated state, non-photochemical fluorescence quenching in actinic light was measured as $Q_{cn} = (F_m-F_m')/F_m$. Epidermal cells (except for stomatal guard cells) normally do not contain chloroplasts, therefore only in mesophyll protoplasts chlorophyll fluorescence was measured.

The protocol used for those measurements had a total length of 300 s and is described in detail in Küpper et al. (2007a). The time-averaged intensity of the measuring light was 4.7 µmol photons.m^{-2}s^{-1}, which is below the threshold causing actinic effects in *Thlaspi*. The applied light intensity for the actinic light was 500 µmol photons.m^{-2}s^{-1}.

A filter for blue excitation light was used (wavelength 410-500 nm; from Photon Systems Instruments, Brno, Czech Republic, www.psi.cz), the dichroic mirror had an edge at 510 nm (from Photon Systems Instruments), and fluorescence was detected between 650 nm and 700 nm (filter from Photon Systems Instruments).

Viability and membrane integrity test

Integrity of cellular membranes and viability of epidermal cells was tested with fluorescein diacetate (FDA, purchased from Invitrogen, department Molecular Probes, Eugene, Oregon, USA, www.invitrogen.com). This dye becomes fluorescent when the acetate groups are cleaved off by esterases inside the cell, yielding fluorescein. If the membranes are intact, diffusion of the fluorescein (in contrast to FDA) is slow, so that cells remain fluorescent for several hours with only a gradual decrease of the signal. If membranes become leaky, fluorescein leaves the cells quickly.

For labelling, 20 µM of FDA were added to the medium reservoir. FDA was detected with excitation at 490-510 nm (AHF HQ 500/20), dichroic mirror at 515 nm (AHF Q 515 LP) and emission at 520-550 nm (AHF 535/30).

Dye calibration

For calibrating the relation between the fluorescence yield of Rhod5N and the concentration of cadmium in the medium, its tripotassium salt was used, purchased from Invitrogen (Department Molecular Probes, Eugene, Oregon, USA, www.invitrogen.com). This Rhod5N tripotassium salt is already hydrolysed and binds metals without passing a cell membrane.

The procedure with the hydrolysed dye was exactly the same as with the "normal" Rhod5N AM. It was dissolved in DMSO first, then IM was added to reach a final concentration of 200 µM Rhod5N tripotassium salt. This solution was mixed 1:1 with cadmium (15 concentrations from 0 nM to 5000 µM), zinc (9 concentrations from 10 nM to 500 mM) and calcium (8 concentrations from 5 µM to 2 M), all of them dissolved in IM. Each sample was pipetted independently at least three times. The mixture of dye and metal in IM was applied to a counting chamber. One minute after mixing it was measured with the same irradiance, objective and filter set as used for the cadmium uptake measurements. A calibration curve was fitted to these data (see Fig. 1). With this curve, it was possible to calculate how much cadmium was taken up by the cell of interest. For testing the effect of competition of weak ligands such as organic acids (as they are known top bind Cd in *Thlaspi*, Küpper et al. 2004), we made a second calibration with addition of 10 mM sodium citrate (see supplemental figure 1).

As the vacuole has an acidic pH while the cytosol is usually close to neutral, the fluorescence yield of the dye could be somewhat different in these compartments (slightly lower in the vacuole). This might lead to underestimation of absolute concentrations in the vacuole, but it does not affect our conclusions about kinetics because there is no reason to assume that the pH would change during the uptake.

Analysis of data measured with the FKM

The whole protoplast and a section of the background were selected from the measured images or two-dimensional maps of parameters (Rhod5N fluorescence or photosynthetic parameters like F_v/F_m) calculated from them. For those two objects, the signal value was calculated and exported out of the program ("Fluorcam 7" from Photon Systems Instruments, Brno, Czech Republic, www.psi.cz.). After export, the background was subtracted from the selected object in order to remove the influence of stray light, and the raw fluorescence values were converted to Cd concentrations using the equation of the calibration curve (Fig. 1). Also images of protoplasts during a measurement were exported; for images of Cd distribution the same calibrated recalculation as for the numerical data was applied in order to obtain quantitative maps of Cd concentrations instead of non-quantifiable fluorescence images. For creating graphs and curves, the program Origin 8.0 Professional (OriginLab, Northampton, Massachusetts, USA) was used. The uptake rates were calculated in this program for each individual cell. A linear regression of the graph was made in the region shortly after the uptake had started for as long as it remained approximately linear; for an example see Fig.°2.

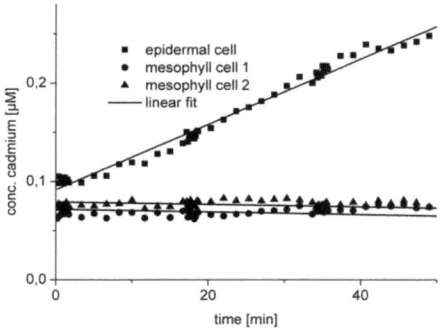

Fig 2. Measurement of two mesophyll and one epidermal cell: time vs. apparent concentration of cadmium, linear fits for uptake rates after background subtraction and subsequent conversion of the Rhod5N-signal into Cd concentrations via the Rhod5N calibration as shown in Fig. 1.

Statistical analysis

Statistical analyses such as ANOVA's were performed in SigmaPlot 11 (SPSS Science, www.spss.com). For statistics, the uptake rates into storage cells were compared to those into standard sized epidermal cells as well as mesophyll cells.

Results

In this study, we introduced a quantitative single-cell *in vivo* measurement of Cd uptake and sequestration, and used it for studying metal uptake kinetics into individual protoplasts of the Cd/Zn hyperaccumulator *T. caerulescens*.

Calibration of the fluorescent dye

For making the assay of Cd-uptake quantitative, the fluorescence response of the dye that was used for the measurement of cadmium uptake into cells, Rhod5N-AM, had to be calibrated for a large range of cadmium concentrations, as shown in Fig. 1. We measured it for 3 orders of magnitude, covering the whole range of cadmium concentrations in the cells between non-accumulator levels and concentrations know from long-term cadmium accumulation.

Furthermore, the fluorescence response of Rhod5N had to be calibrated not only for cadmium. Because of the high abundance of Ca and Zn in the cells, combined with their chemical similarity to Cd, the dye had to be calibrated for zinc and calcium as well to make sure that the detected signals mainly came from cadmium and not from zinc or calcium. Fig.°1 shows that the curve for cadmium saturated at a much lower concentration than for zinc or calcium. Obviously, Rhod5N-AM has a very good selectivity for cadmium compared to zinc or calcium, confirming results of Soibinet et al. (2008). For quantification of cadmium uptake measurements, a sigmoidal fit was calculated in the physiological range of our short-term uptake experiments (resulting in concentrations lower than 10 µM, Fig. 1). To obtain half-saturation of the maximal, Cd-induced dye fluorescence, 54 µM cadmium, 1,300 µM zinc and 253,000 µM calcium were needed. The extremely large difference between Cd and the other metals was also partially caused by the fact that even at saturating metal concentrations the Rhod5N complexes with Ca^{2+} and Zn^{2+} were far less fluorescent than the Rhod5N complex with Cd^{2+} (Fig. 1A). The different half saturation for cadmium also means that not only free but also weakly bound Cd^{2+} should be efficiently detected, which we tested with the addition of citrate (supplemental Fig. 1). This test showed no competition for Cd binding to the dye by the citrate ligands, which would manifest itself in a shift of the half-saturation concentration and possibly a lower maximal fluorescence, but neither effect was observed. In contrast to Cd, for Ca^{2+} and Zn^{2+} a multitude of other potential ligands in cells (even organic acids included) will be binding stronger than Rhod5N, making the dye even more selective in cells than shown by our calibration.

The curve (metal concentration plotted against relative fluorescence) was difficult to fit in the region of higher cadmium concentrations as it did not follow a simple sigmoidal curve. Therefore, only the part with lower concentrations, from 0 μM to 5 μM, was fitted using a sigmoidal function (Fig. 1B). This region was the most interesting part of the curve, because it represented the physiological range of cadmium in the cells treated with cadmium for these few hours of an uptake measurement. In hyperaccumulator plants grown for several months on a metal for which they have hyperaccumulation capabilities (e.g. Cd+Zn, but not Cu for *T. caerulescens*), the concentrations of this metal in the vacuoles of epidermal cells reach much higher concentrations Küpper et al. 1999, 2000, 2001).

In addition to Rhod5N we tried to use another dye for cadmium and lead, "Leadmium", which was recently developed by Invitrogen and used by Lu et al. (2008). Unfortunately, cells showed a very intense fluorescent signal in the spectral emission region of the dye already before the addition of cadmium, indicating a strong autofluorescence (supplemental Fig. 2).

Distribution of cell types and sizes in a typical protoplast sample

As described in more detail in the introduction, from previous studies it was known that heavy metal accumulation in hyperaccumulating Brassicaceae mainly occurs in large epidermal cells. Therefore, before applying statistics to our data we grouped the epidermal cells by their size. Figure 3 shows a histogram of the distribution of cell types and sizes obtained from a typical sample (whole leaf digestion) in the measuring chamber. It is obvious that the size distribution of epidermal cells was much wider compared to mesophyll cells, where all cells were of rather similar size. As far as it is possible to judge from electron and confocal microscopy pictures, this size distribution pattern corresponds to the pattern observed in earlier studies on *T. caerulescens* Ganges (Küpper and Kochian (2010) and Prayon (Küpper et al. 1999), Frey et al. 2000) ecotypes as well as *T. goesingense* (Küpper et al. 2001). All protoplasts from the upper epidermis with a diameter smaller than 60 μm were named "standard epidermal cells". Cells larger than 60μm were only present in the epidermis, not at all in the mesophyll. As these very large epidermal cells were clearly separated from all other cells, in the following we refer to them as "storage cells", knowing that metals are stored in such especially large epidermal cells (see introduction). For a sample image in transmitted light see Fig. 4.

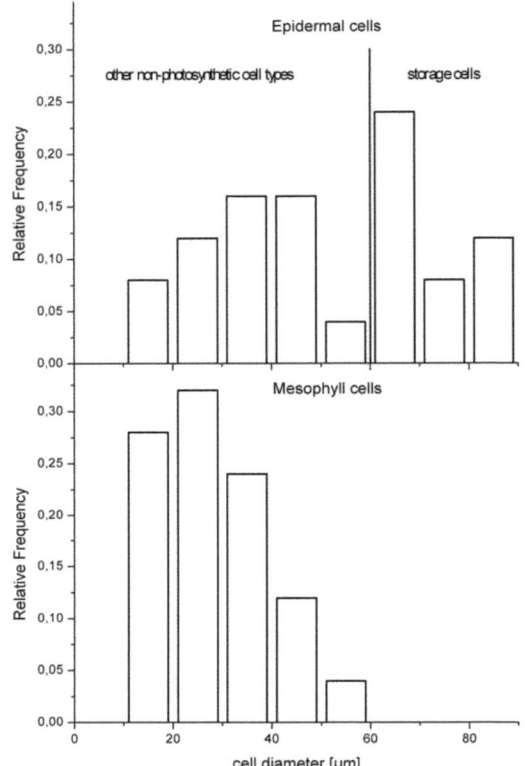

Fig 3. Types and sizes of cells obtained from a typical protoplast preparation from a young-mature (i.e. just after reaching its full size) leaf of *Thlaspi caerulescens* (Ganges ecotype).

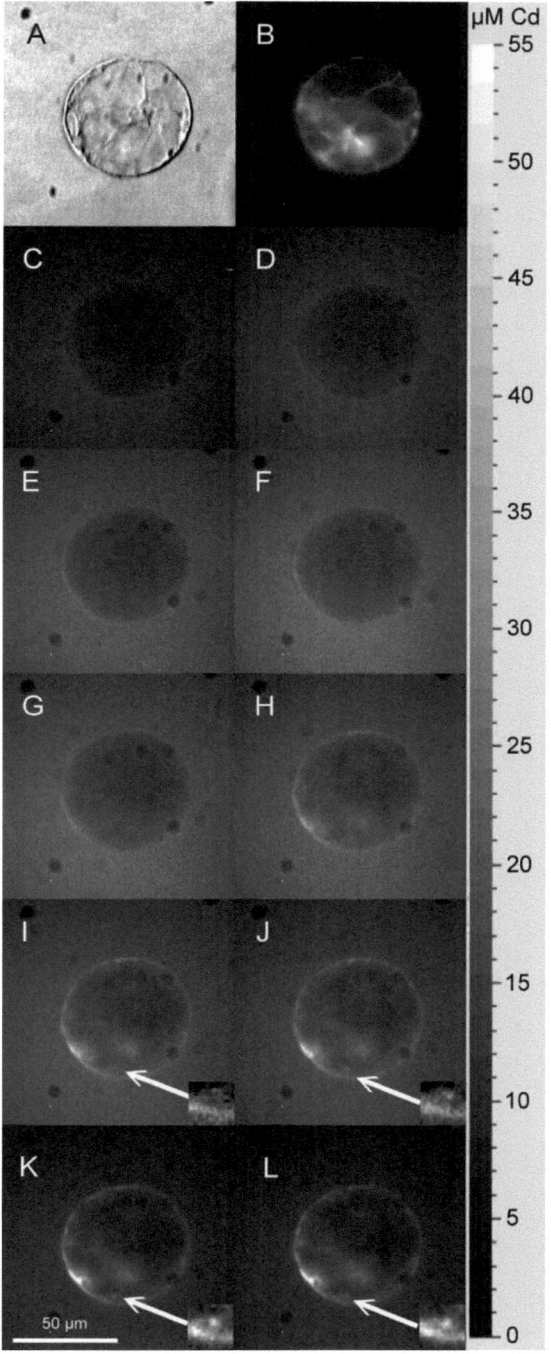

Fig 4. Early stages of Cd-uptake and sequestration in an epidermal storage protoplast. (A) transmitted light image, (B) vitality test via FDA fluorescence, (C) – (L): time series of Cd concentration maps after addition of 1 μM Cd^{2+}. (C) 10 s, (D) 30 s, (E) 100 s, (F) 200s, (G) 300 s, (H) 400 s, (I) 500 s, (J) 600s, (K) 900 s, (L) 1000 s. In the small inset pictures the scale is stretched to an amplitude of 12 μM for contrast enhancement. The arrows point at Cd transporting vesicles.

Isolation of vital mesophyll and epidermal protoplasts of *Thlaspi caerulescens*

In this work we established a protocol for the isolation of vital epidermal protoplasts based on protocols of Coleman et al. (1997) and Ferimazova et al. (2002). After removing the lower epidermis, the Petri dishes containing the leaves were not placed on a shaker but kept on a temperature controlled (23°C) metal plate without shaking as it turned out that shaking destroys many epidermal cells. Osmolarity of the isolation medium turned out to be most critical for the survival of the epidermal storage cells. In the Petri dishes, the protoplasts obtained in this way stayed vital for at least 24 hrs. In the measuring chamber with repeated measurements they survived well for several hours, as tested by repeatedly measuring FDA of all cells and photosynthetic performance of mesophyll cells (Fig.°5A). The most difficult work was to get a sufficient amount of the large epidermal protoplasts (storage cells with a diameter >60 µM), because these are the most fragile cells so that they easily rupture e.g. during the washing steps, the mixing with agarose or the mounting of cellophane on the chamber (see methods). Nevertheless, in the end we obtained a sufficient number of cells of each type for the analysis of Cd uptake. An example of the vitality test with FDA and maps of Cd-uptake measured via the calibrated Rhod5N fluorescence response in a storage cell is shown in Fig. 4.

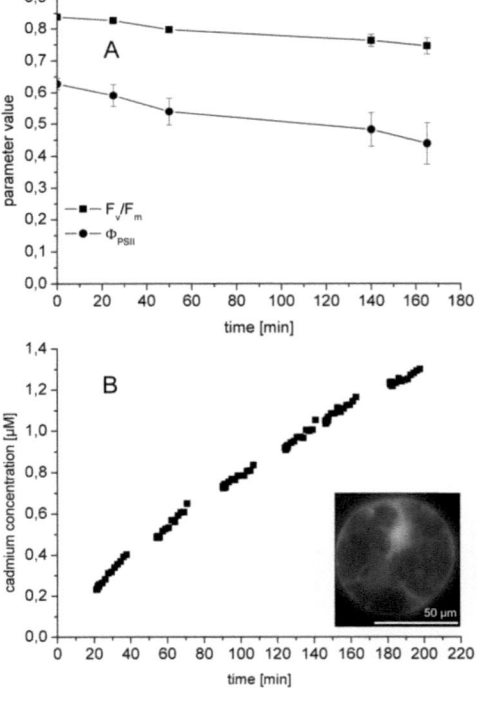

Fig. 5. Tests of the long-term viability of the cells under measuring conditions. **(A)** Measurement of photosynthetic parameters in mesophyll cells over a time period of 170min. **(B)** Measurement of an epidermal storage cell. Time vs. apparent concentration of cadmium, example of an FDA viability / membrane integrity test for a living storage cell during cadmium uptake.

Physiological results

The first important result was the successful observation of cadmium uptake into living protoplasts via the fluorescent dye. It usually started about 2-15 min after addition of Cd to the medium reservoir that supplies the medium to the chamber (see Küpper et al. 2008). By *in vitro* experiments using the potassium salt of Rhod5N in the counting chamber that was also used for the dye calibration, we found that this delay was due to the diffusion time of Cd through the layer of 0.16% agarose in which the cells were embedded (data not shown). Once uptake had started, it could often be followed in continuous measurements for at least one hour, sometimes up to 3.5 h (Fig.°5B). By recalculating the measured fluorescence kinetics (after background subtraction) with the dye calibration, we could collect quantitative uptake data that allowed for separating between two different phases of the uptake, and to characterise the Cd uptake of different cell types, as described in detail below.

Appearance of a cytoplasmic ring

A few minutes after the application of cadmium to the medium, a bright ring of Rhod5N fluorescence became visible near the plasma membrane (Figs. 4, 6). Comparison with transmitted light pictures of the same cell revealed that the ring was located on the inner side of the plasma membrane and outside the tonoplast, so that it could be clearly identified as the cytoplasm (Fig. 4). The cytoplasmic ring appeared in all measured cells and in all cell types. This ring was visible for some time (exact time depending on the individual cell), then the Cd-signal in the cytoplasm saturated and slowly the whole cell became filled with cadmium with highest apparent concentrations finally reached in the vacuole (Fig. 6). All non-photosynthetic epidermal cells have large vacuoles, and in particular in the epidermal storage cells the vacuole fills almost the whole cell. In cases where cells were loaded with cadmium before the incubation with dye (i.e. the opposite of the normal procedure), no cytoplasmic ring appeared, but Rhod5N fluorescence was strongest in the centre (= vacuole) of the cell (see Fig.°6D), so that it can be excluded that the cytoplasmic ring was an artefact of dye distribution (too short pre-loading of the cells with the dye).

In several metal storage cells we observed vesicle-like Cd-rich structures during the time from appearance of the cytoplasmic ring until the vacuole filled with Cd (see arrows and small inset maps in Fig. 4). These vesicles appeared and moved in the cytoplasm, but since they did not remain in focus for long enough it was not possible to determine where they originated or where they migrated to.

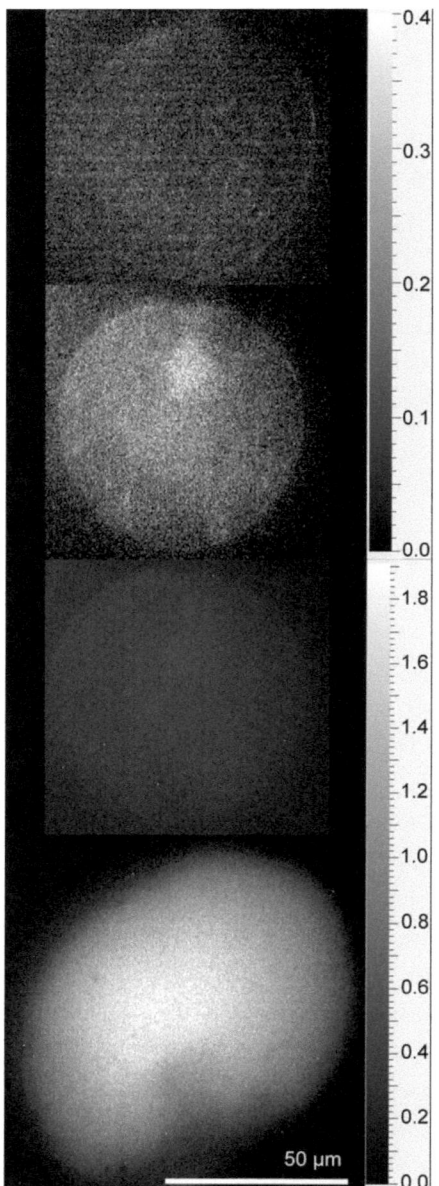

Fig. 6. Phases of Cd-uptake kinetics in epidermal storage protoplasts: maps of apparent Cd concentrations during the different phases of Cd uptake and sequestration. Maps (A)-(C) are of the same protoplasts, which was incubated with 10 µM cadmium in the measuring chamber. Map (D) is from a different preparation, where over night Cd uptake was tested with about 100 nM Cd (lower concentration used for enhancing long-term viability of the cells). The scales of maps (A) and (B) are adjusted for optimal visibility of the short-term uptake, the scales of maps (C) and (D) are adjusted for optimal comparison of the apparent concentrations reached after short term versus over night incubation.

Cadmium uptake into mesophyll cells

Both Chl fluorescence and the light absorption by chlorophyll in the chloroplasts were used to differentiate them from epidermal cells. The latter were not filled with chloroplasts and therefore showed no Chl fluorescence but were much brighter in transmitted light. All of the measured mesophyll cells took up cadmium. However, the uptake rates of individual mesophyll cells differed strongly from each other (supplemental Table 1). Trying to correlate size and uptake rate of mesophyll cells, no clear trend was observed, neither in the velocity of uptake nor in the apparent concentration which was reached finally in the individual cell (not shown). For a combined measurement of epidermal cells together with a cell from the mesophyll, see Fig. 2.

For the mesophyll cells we observed a few cases where the cells took up a lot of cadmium in a very short time, usually occurring at the end of a long measurement where no clear uptake was visible. This sudden rapid uptake was usually followed by rupture of the cell. Therefore, we did not use these rare cases for statistics but strongly suspect a defective cell membrane as a cause.

Cadmium uptake into epidermal cells

Again all measured cells showed an uptake of cadmium, but at least in larger epidermal cells uptake rates were higher than in mesophyll cells. For a combined measurement of epidermal cells together with a cell from the mesophyll, see Fig. 2. This difference was due to a strong size dependence of uptake rates in epidermal cells. One of the important findings of this work was that uptake rates for cadmium in epidermal cells of different sizes strongly differ from each other (see Table 1). While the ANOVA, which was calculated for all uptake rates of all mesophyll and epidermal cells measured, showed no significant ($P = 0.05$) difference between mesophyll cells and "standard" (small) epidermal cells, the epidermal storage cells had almost 14 times higher Cd uptake rates (for statistics see Table°1) than the mesophyll cells. Furthermore, the Cd-containing vesicles (see above) were only found in storage cells.

DISCUSSION

In this study, we were able to introduce a quantitative single-cell *in vivo* measurement of Cd uptake and sequestration, and to use it for solving two key questions of Cd accumulation in leaves, which are the main metal storage sites of hyperaccumulator plants: 1. Is the preferential heavy metal accumulation in large epidermal storage cells compared to other cell types due to differences in active metal transport?
2. Which is the time limiting step in metal sequestration into large epidermal cells?

Quantitative *in vivo* measurement of cadmium uptake kinetics

All earlier studies with metal-specific fluorescent dyes in plants, and to our knowledge also in other organisms, were carried out without proper (i.e. under measuring conditions) calibration of the fluorescence response versus the dye concentration. This calibration turned out to be essential for any quantitative work with such dyes because of the extremely non-linear response of the dye. The strongly sigmoidal response of the dye at increasing Cd concentrations can be interpreted as an indicator of cooperative binding of Cd to the dye molecules, e.g. via aggregation of the Cd-dye complexes. This could also explain why it was different in our calibration compared to the data of Soibinet et al. (2008) that were obtained in another buffer. Therefore, previous studies using metal-specific fluorescent dyes that unfortunately did not use any calibration produced only qualitative results. Additionally, it was important to show that neither zinc nor calcium, which both bind to Rhod5N as well and are present in all living cells, disturb the quantification of cadmium. As it was used by Lu et al. (2008) for looking at Cd uptake in plant roots, we furthermore tested another dye for cadmium and lead, "Leadmium". However, the very strong autofluorescence in the specific emission region of the dye made Cd uptake measurements with this dye impossible. Furthermore, as its only application to plant cells (Lu et al. 2008) did not include any controls and we observed strong autofluorescence in the spectral emission range of leadmium, the conclusions drawn in that study have to be questioned. This autofluorescence in the green spectral region is a well known phenomenon observed for plants cells in contrast to animal cells, which do not have that many interfering substances in their cytoplasm.

Insights into cellular cadmium accumulation and sequestration

Before this study, several attempts had been made to isolate vital protoplasts from the epidermis of *T. caerulescens* but the methods that were used did not work for this purpose (Cosio, Martinoia & Keller 2004, Cosio et al. 2005, Ma et al. 2005). In this work we established a protocol for the isolation of vital epidermal protoplasts using protocols of Coleman et al. (1997) and Ferimazova et al. (2002).

The Cd uptake rates and final apparent Cd concentrations in the large epidermal protoplasts (metal storage cells) differed a lot, but they were in all cases higher compared to the standard epidermal cells and mesophyll cells. This was an important finding of this study, because it shows that the differences in cellular metal concentrations observed after long-term hyperaccumulation of metals (e.g. Küpper, Zhao & McGrath 1999, Küpper et al. 2001; Frey et al. 2000; Bidwell et al. 2004; Bhatia et al. 2004; Broadhurst et al. 2004; Cosio et al. 2005) are due to differences in active transport of the cells, as opposed to differences in cell wall properties or transpiration stream. If differences in cell wall properties or transpiration stream had caused the differential accumulation of epidermal cells, preparing protoplasts from them (i.e. taking them out of their tissue context and removing the cell walls) would have eliminated these differences. Cd sequestration into storage cells makes sense as those cells use their vacuole to store metal safely and without harm to sensitive enzymes, and they do not contain a photosynthetic apparatus that would be a sensitive target of Cd-toxicity.

A recent study on the cellular distribution and regulation of gene expression levels (Küpper and Kochian, 2010) indicates proteins that may be candidates for causing this difference in metal accumulation between epidermal cells: TcZNT5 and TcMTP1 were highly expressed in epidermal storage cells, much higher than in other cell types. The *Arabidopsis thaliana* homologue of the latter protein, AtMTP1, a member of the cation diffusion facilitator family of heavy metal transporters, had already previously been shown to mediate Zn detoxification and leaf Zn accumulation (Desbrosses-Fonrouge et al. 2005). Due to the chemical similarity of cadmium and zinc, transporters designed for Zn transport Cd as well for purely chemical reasons (details below). While MTP1 is known to be localised in the vacuolar (tonoplast) membrane (Desbrosses-Fonrouge et al. 2005), the ZIP-family of transporters, to which ZNT5 belongs, is generally localised in the plasma membrane (Review by Guerinot, 2000). Therefore, the results of the current study shed a new light on the storage cell expression of ZNT5 found previously (Küpper and Kochian, 2010). Our statistics showing enhanced uptake rates in the storage cells is based on accumulation times in the range of 1-2 hours, where accumulation mainly occurred in the cytoplasm. Therefore, it can be concluded that the higher uptake rates in epidermal storage cells compared to other cell types are partially caused by differences in transport over the plasma membrane, and its expression pattern suggests that ZNT5 is involved in this phenomenon.

In the current case the investigated metal was cadmium, while in the studies of metal distribution after long-term accumulation zinc and nickel were used (Küpper et al. 1999, 2001; Frey et al. 2001). But as the ecotype Ganges of *T. caerulescens* accumulates both cadmium and zinc, it could be expected that both metals are stored in a similar way, especially in view of the similar chemical properties of the two metals. Indeed, a later study of Cd accumulation in this species showed the sequestration in epidermal storage cells as well (Cosio et al. 2005), despite problems with cell rupture during the fractionation that led to increased cell wall binding in this study. Further, accumulation in large epidermal cells was even observed for nickel in various plants (Küpper et al. 2001; Psaras et al. 2000, Bhatia et al. 2004), indicating that it is a rather general strategy of hyperaccumulators. Similarities in transport for different hyperaccumulated metals originate, most likely, at the protein level. Known Zn transporters have an affinity both for zinc and cadmium (e.g. the CPx-ATPase TcHMA4: Papoyan & Kochian 2004, Parameswaran et al. 2007), which also has to be expected from the chemical point of view (Irving-Williams series). However, an uptake competition study revealed no inhibition of Cd uptake by Zn competition (Lombi et al. 2000). This indicated that the rate limiting step of uptake in *T. caerulescens* (Ganges) involves different transporters for both metals, at least in addition to proteins transporting both metals.

The appearance of the transient Cd-accumulation in the cytoplasm in almost all measured cells and in all three of the analysed cell types clearly showed that the transport into the vacuole is the rate -limiting step in cadmium uptake into protoplasts. This means that at the beginning of metal uptake the metal accumulates in the cytoplasm, because the sequestration into the vacuole is slower than the uptake over the plasma membrane. This could be due to a smaller number or lower turnover rates of metal transporters situated in the vacuolar membrane compared to those in the plasma membrane. When a certain metal level is reached, the cytoplasmic accumulation saturates, possibly by feedback inhibition of the transporters over the plasma membrane. The observation that longer-term accumulation finally leads to filling of the vacuole to much higher apparent concentrations than those reached in the cytoplasm (in previous studies, high millimolar concentrations were found, Küpper et al. 1999, 2000, 2001) indicates that the vacuolar transporters are slower or fewer than those in the plasma membrane, but have a higher translocation efficiency. The observation of Cd-containing vesicles in the time period of the cytoplasmic ring indicates that besides direct transport over the tonoplast membrane also vesicular transport may play a role in vacuolar sequestration (or in efflux from the cell). This is an interesting topic for further studies. A limitation of vacuolar accumulation by phytochelatin synthesis is very unlikely, because inhibition of phytochelatin synthase does not affect Cd resistance of *T. caerulescens* (Schat et al. 2002), phytochelatin levels in this species are even lower than in related non-accumulator species (Ebbs et al. 2002), and Cd is not stored in association with sulphur ligands like phytochelatins (Küpper et al. 2004). In contrast, very many recent studies have demonstrated that the hyperaccumulation phenotype is related to enhanced expression of metal transporters (see introduction). So it remains most likely that the now observed limitation of Cd accumulation by vacuolar sequestration is caused by limiting availability of metal transporters.

Candidate proteins for the rate-limiting step of vacuolar sequestration are, according to previous expression studies, *Thlaspi caerulescens* homologues of HMA3, MHX, MTP1 and ZIF1. HMA3, a CPX- (= P_{1B})-type heavy metal ATPase, was found to mediate leaf vacuolar storage of Cd, Co, Pb and Zn in *Arabidopsis thaliana* (Morel et al. 2009). MHX, a homolog of an *Arabidopsis thaliana* vacuolar metal (Fe, Mg, Zn) versus proton exchanger and member of the cation diffusion facilitator (CDF) protein family, was found to be highly expressed in the leaf vacuolar membrane of *Arabidopsis halleri* (Elbaz et al. 2006). MTP1, another member of the CDF family, was shown to be a Zn transporter in the vacuolar membrane of *A. thaliana*, and was furthermore shown to mediate Zn detoxification and Zn accumulation in the leaves (Desbrosses-Fonrouge et al. 2005). Finally, ZIF1, a member of major facilitator protein superfamily, was found to be expressed in *A. thaliana* leaf vacuolar membranes, and also this protein influenced Zn tolerance and accumulation of the plants (Haydon and Cobbett, 2007). The finding of the current study that the transport from thy cytoplasm to the vacuole is the rate-limiting step of accumulation now explains why overexpression of the aforementioned proteins did not only lead to enhanced metal resistance, but metal accumulation. As this enhanced metal accumulation was observed on the whole-plant level, it now seems likely the transport over the vacuolar membrane in the metal storage cells of the leaf epidermis is not only the rate-limiting step of metal hyperaccumulation in leaf cells as analysed now, but that it is an important driving force behind the complex phenomenon of metal hyperaccumulation.

ACKNOWLEDGEMENTS

The authors would like to thank Gabriela Lutz for measuring a few of the cells during an advanced training course in plant physiology, and Aravind Parameswaran for help with the maintenance of some of the plants. Additionally, we would like to thank Pavel Korabečny and Martin Trtílek for special modifications of the FluorCam software that were required for the recording and quantification of the data in this study. This study was financially supported by grants of the Landesstiftung Baden-Württemberg, Universität Konstanz (AFF) and the Fonds of the Chemical Industry (FCI).

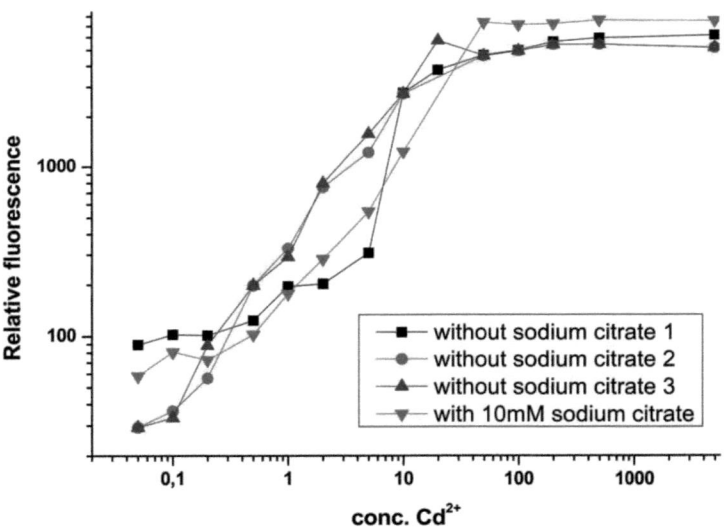

Supplemental Fig. 1 Calibration of Rhod5N in the presence of 10mM Sodium citrate and without citrate addition.

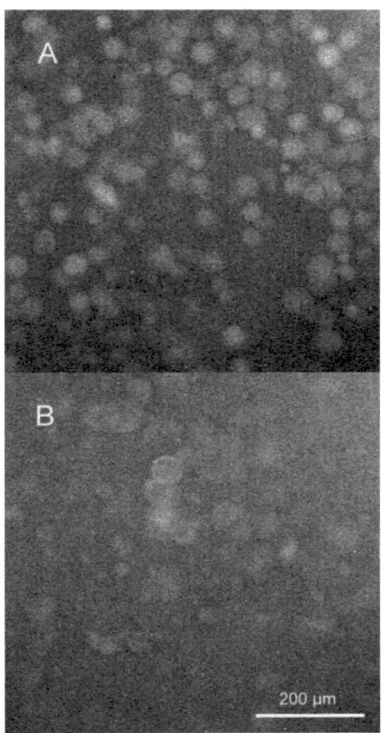

Supplemental Fig. 2 A: Fluorescence signal of protoplasts incubated with Leadmium and no addition of cadmium, B: Fluorescence siganl of protoplasts incubated with Leadmium after the addition of cadmium.

Table 1: Statistics of Cd-uptake kinetics in individual protoplasts. Each measurement (protoplast) is from an independent protoplast preparation, i.e. a true replicate. Dependent Variable: Cd uptake rate ($\mu M\ Cd^{2+}.min^{-1}$)

Group Name	N	Mean	Std Dev	SEM
mesophyll protoplast	9	0.00117	0.00205	0.000684
small epidermal protoplast	6	0.00256	0.00200	0.000816
epidermal storage cells	4	0.0159	0.0133	0.00664

Source of Variation	DF	SS	MS	F	P
Between Groups	2	0.000637	0.000319	8.741	0.003
Residual	16	0.000583	0.0000364		
Total	18	0.00122			

The differences in the mean values among the treatment groups are greater than would be expected by chance; there is a statistically significant difference ($P = 0.003$).

Power of performed test with alpha = 0.050: 0.911

All Pairwise Multiple Comparison Procedures (Holm-Sidak method): Overall significance level = 0.05. Comparisons for factor: cell type

Comparison	Diff of Means	t	Unadjusted P	Critical Level	Significant?
storage epi vs. meso p.	0.0147	4.047	<0.001	0.017	Yes
storage epi vs. small epi p.	0.0133	3.412	0.004	0.025	Yes
small epi pr vs. meso p.	0.00139	0.436	0.669	0.050	No

2.3. A native Zn/Cd pumping P_{1B} ATPase from natural overexpression in a hyperaccumulator plant

Aravind Parameswaran[a,b,*], Barbara Leitenmaier[a,*], Mingjie Yang[a,c], Peter M.H. Kroneck[a], Wolfram Welte[a], Gabriela Lutz[a], Ashot Papoyan[d], Leon V. Kochian[d] and Hendrik Küpper[a,e]

[a] Universität Konstanz, Fachbereich Biologie, 78457 Konstanz, Germany

[b] current address: Novartis Healthcare Pvt. Ltd, NICCI India, ILabs HTC PVt. Ltd, HITEC City, Hyderabad 500 081, India.

[c] Fudan University, Institute of Modern Physics, Shanghai 200433, PR China

[d] US Plant, Soil & Nutrition Laboratory, USDA-ARS, Cornell University, Ithaca, NY 14853, USA

[e] Faculty of Biological Sciences and Institute of Physical Biology, University of South Bohemia, Branišovská 31, CZ-370 05 České Budejovice, Czech Republic

* the first two authors have contributed equally to this study

published in 2007 in BBRC 364: 51-56

Abstract

TcHMA4 is a P_{1B}-type ATPase that is highly expressed in the Cd/Zn hyperaccumulator plant *Thlaspi caerulescens* and contains a C-terminal 9-histidine repeat. After isolation from roots, we purified TcHMA4 protein via metal affinity chromatography. The purified protein exhibited Cd- and Zn-activated ATPase activity after reconstitution into lipid vesicles, showing that it was in its native state. Gels of crude root extract and of the purified protein revealed TcHMA4-specific bands of about 50 and 60 kDa, respectively, while the *TcHMA4* mRNA predicts a single protein with a size of 128 kDa. This indicates the occurrence of post-translational processing; the properties of the two bands were characterised by their activity and binding properties.

Keywords: Cadmium, heavy metals, metal activation, P_{1B} type ATPase, TcHMA4, zinc

Introduction

Metal hyperaccumulating plant species not only tolerate high amounts of heavy metals like zinc or cadmium in the environment, but also take them up actively and accumulate them up to several percent of their shoots dry mass. This ability makes them attractive for cleaning up metal-contaminated soils (phytoremediation: Salt et al., 1995, Raskin et al., 1997). Many studies have been conducted to reveal mechanisms of heavy metal hyperaccumulation. Nonetheless, very little is known about mechanisms and proteins involved in transporting hyperaccumulated heavy metals from soil via the roots and stems into their storage sites. These are mainly large epidermal storage cells in the leaves (Küpper et al., 1999, Frey et al., 2000). The transport proteins involved in hyperaccumulation might belong to different families such as the CPx-type ATPases (P_{1B} ATPases), ZIPs, Nramps, CDFs and CAXs (Hall and Williams, 2003, Williams and Mills, 2005, Küpper and Kroneck, 2007). P_{1B} ATPases are a subfamily of the P-type ATPase superfamily, a group of ubiquitous membrane proteins that use ATP to pump cations across membranes against their electrochemical gradient. They are found in a wide range of organisms including bacteria, yeast, plants and animals including humans (Solioz and Vulpe, 1996). Plant P-type ATPases have eight to twelve transmembrane (TM) domains, N and C termini exposed to the cytoplasm, and a central cytoplasmic domain which harbours phosphorylation and ATP binding sites (Palmgren and Harper, 1999, Axelsen and Palmgren, 2001). Surprisingly, while many genes encoding heavy metal transporters have been identified, very few of the proteins encoded by these genes have been isolated, purified and biochemically characterised, although this is necessary to understand their mechanisms of metal transport. No crystal structures for any heavy metal transporter of the ZIP,

CDF and CPx-ATPase families in eukaryotes have been described, and only peptide subdomains of any of these transport proteins have been expressed and characterised in bacteria (Verret et al., 2005, Wunderli-Ye and Solioz, 2001). One reason for the lack of such studies are most likely the well-known difficulties with heterologous overexpression of membrane proteins, in particular those that transport potentially toxic substances like heavy metals. Membrane proteins expressed in bacteria or yeast systems often exhibit incorrect folding, and furthermore it is difficult to judge if the target protein is post-translationally modified as it would be if expressed in its natural system. Expression systems like bacteria and yeast are not able to tolerate the high heavy metal levels which can be reached in the cells when a metal transporter is overexpressed in such systems.

Hyperaccumulators offer a way to circumvent those problems. First, these plants have a strongly elevated expression of metal transporters (Pence et al., 2000, Assuncao et al., 2001, Becher et al., 2004, Papoyan and Kochian, 2004, Weber et al., 2004). Second, the protein studied here, TcHMA4, is a CPx-type Cd/Zn-ATPase (1185 amino acids in length). It has a natural C-terminal sequence of 9 His residues (Papoyan and Kochian, 2004, Bernard et al., 2004)) which can be used as a natural tag for its purification by metal affinity chromatography. Finally, isolating the protein in this way will reveal possible post-translational modifications. Therefore, we applied this strategy and systematically developed a protocol for purifying TcHMA4 to homogeneity in active state.

Materials and Methods

Plant material, culture media and culture conditions

Thlaspi caerulescens J.&C. Presl (Ganges population from southern France) was germinated and grown as described in detail in Küpper et al., 2007, using 100 µM $ZnSO_4$ in the nutrient solution for Zn-replete but non-inhibitory conditions. The nutrient solution was aerated and was exchanged continuously at a flow rate of 1700 ml/d^{-1} per pot (i.e. 250 $ml.d^{-1}$ per plant)

Chemicals

The chemicals were purchased from the following manufacturers:
- chemicals used in the buffers for isolation and solubilisation: Merck, Germany
- detergents: n-dodecyl-β-maltoside (DDM): Carl Roth, Germany; N,N-dimethyldodecylamine N-oxide (LDAO): Fluka, Germany; n-octyl-β-D-glucopyranoside and n-dodecyl-N,N-dimethyl-ammonio-3-propane-sulfonate (zwittergent3-12): Anatrace, USA)
- protease inhibitors: PMSF and PefaBloc: Fluka, Germany; protease inhibitor cocktail VI: Merck, Germany; protease inhibitor cocktail tablets "complete", EDTA-free: Roche, Switzerland

- antioxidants and chemicals for column buffers: Merck, Germany
- chemicals used for ATPase activity test: Sigma-Aldrich, Germany

Column materials

The following immobilised metal affinity column materials were compared:
- precharged "Protino" Ni-IDA (Macherey-Nagel, Germany).
- precharged "Protino" Ni-TED (Macherey-Nagel, Germany).
- "Talon" resin precharged with Co^{2+} (Takara Bio Europe, France)
- precharged Ni-sepharose "fast flow" (GE Healthcare, UK)
- uncharged His-Bind Fractogel (www.novagen.com), charged before the run with Cd^{2+}, Cu^{2+} or Zn^{2+}

Other materials

centrifugal protein concentrators: Amicon Ultra-15, membrane with 10 kDa exclusion size (Millipore Corporation, USA)

Instruments

Instruments of the following manufacturers were used:
- wheat mill: Jupiter 872 with Messerschmidt stainless steel grinding engine (Messerschmidt Hausgeräte GmbH, Germany)
- ultracentrifuge: Beckmann LE-80 (Beckmann, Germany)

Methods for testing protein purity

SDS-PAGE was conducted as described in Lämmli, 1970, and Western-blotting was done using the Protran BA 85 nitrocellulose membrane (Whatman, UK). The peptide sequence for the primary antibody was generated from the cDNA-deduced amino acid sequence of TcHMA4 by selecting a region of the protein that is unique to this member of the HMA4 family and has a high degree of probability of reacting with the antibody. The peptide sequence, DKEKAKETKLLLASC, is derived from the 3' cytoplasmic tail of the TcHMA4 protein and was provided to Sigma Genosys for antibody production (Sigma-Aldrich, USA). The peptide was conjugated with keyhole limpet hemocyanin, the conjugated peptide was injected subcutaneously into two rabbits and after several boost injections, three bleeds were performed on days 49, 56, and 77 after the injection. A 1:1000 dilution of antiserum from the third bleed was successful in detecting a single band for the TcHMA4 protein in plasma membrane vesicles isolated from *Thlaspi caerulescens* roots and was used in subsequent experiments. The secondary antibody (Anti-rabbit IgG) was purchased from Sigma-Aldrich (Germany).

Protein was determined by the Bicinchoninic acid assay (Smith et al., 1998) and all chemicals for this assay were from Sigma-Aldrich (Germany).

ATPase activity was assayed according to a protocol that was developed as described in Serrano, 1985, and modified for the current study. The liberated phosphate was quantified against phosphate standards. In the first step the ATPase was reconstituted into lipid vesicles. For this purpose the ATPase was diluted to 0.18 mg.ml^{-1} and then mixed in a ratio 1:5 with phosphatidylcholine from soybean. Reconstitution took place for 10 min at 37°C. For the ATPase reaction, the reconstituted protein was transferred into a buffer containing 0.6 M Tris, 2 M NaCl and 0.1 M $MgCl_2$. The reaction was started with the addition of 5 mM ATP and took place for 40 min at 30°C. It was stopped with a solution containing 0.5% SDS, 0.5% NH_4MoO_4 and 2% (V/V) H_2SO_4. The colour was developed with 10% ascorbic acid and the absorbance at 750 nm was read after 2 min.

Results and Discussion
Development of the isolation and purification protocol
Isolation of TcHMA4 protein

Frozen roots were ground to powder using a wheat mill pre-cooled with dry ice, and further ground to a finer powder with liquid nitrogen using a mortar and pestle. Frozen isolation buffer was added to the frozen roots during the grinding and the whole extract was thawed at room temperature. After mixing the suspension, it was centrifuged (246,000xg, 4°C, 1 h). The supernatant containing the cytoplasmic proteins was discarded and the pellet was retained. It was resuspended in solubilisation buffer and continuously stirred at 4° C for 14 hours (standardised by tests) for optimal solubilisation of TcHMA4.

The successful solubilisation of the Zn/Cd ATPase from membranes in a native form was the key step leading to the purification of this protein. Therefore, the composition of the isolation and solubilisation buffers was systematically optimised.

The efficiency of protein solubilisation was first tested with different buffer systems including MES-KOH, NaH_2PO_4-NaOH, sodium citrate-HCl ($C_6H_5NaO_7$-HCl) and sodium acetate-acetic acid (CH_3COONa-$C_2H_4O_2$), of which the phosphate buffer functioned best. Four different detergents were tested (Fig. 1). These included the non-ionic detergents n-octyl-β-D-glucopyranoside and n-dodecyl-β-D-maltoside (DDM), and the zwitterionic detergents N,N-dimethyldodecylamine N-oxide (LDAO) and n-dodecyl-N,N-dimethyl-ammonio-3-propane-sulfonate (zwittergent3-12). As the best results were obtained with DDM, different concentrations of this detergent were tested, showing that >5 mM DDM were needed for solubilisation. Saturation of the amount of isolated protein was reached at about 10 mM (Fig.2). Further tests were conducted to identify the optimal concentrations of NaH_2PO_4 (0-320 mM) and NaCl (0-3.2M). It could be observed that increasing the NaCl concentration up to 2 M proved to be beneficial for extracting a higher amount of the native membrane protein. However, an increase in the concentration of NaH_2PO_4 did not significantly affect the solubilisation. Variations in pH from 5.0 to 8.0 were tested for both the isolation and the solubilisation buffer. It was found that the optimal pH was 6.0, as it led to the highest amount and lowest degradation of isolated protein.

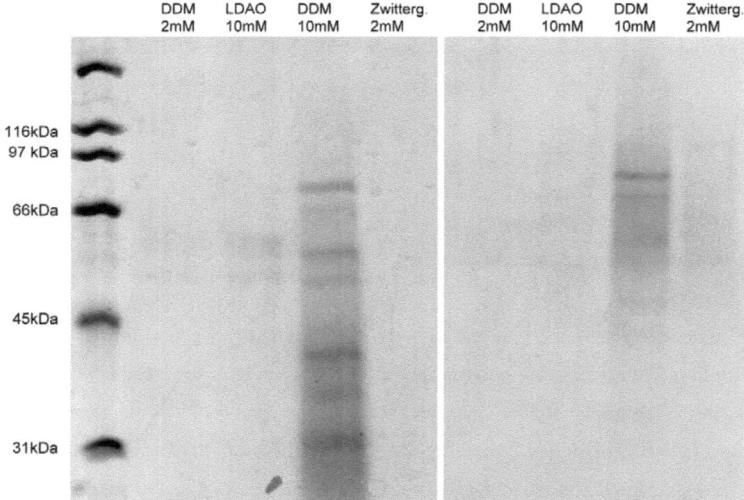

Fig.1 Different detergents in the solubilisation buffer, gel and blot of crude extract. Lanes from left to right: 2 mM DDM, 10mM LDAO, 10mM DDM, 20 mM Zwittergent3-12

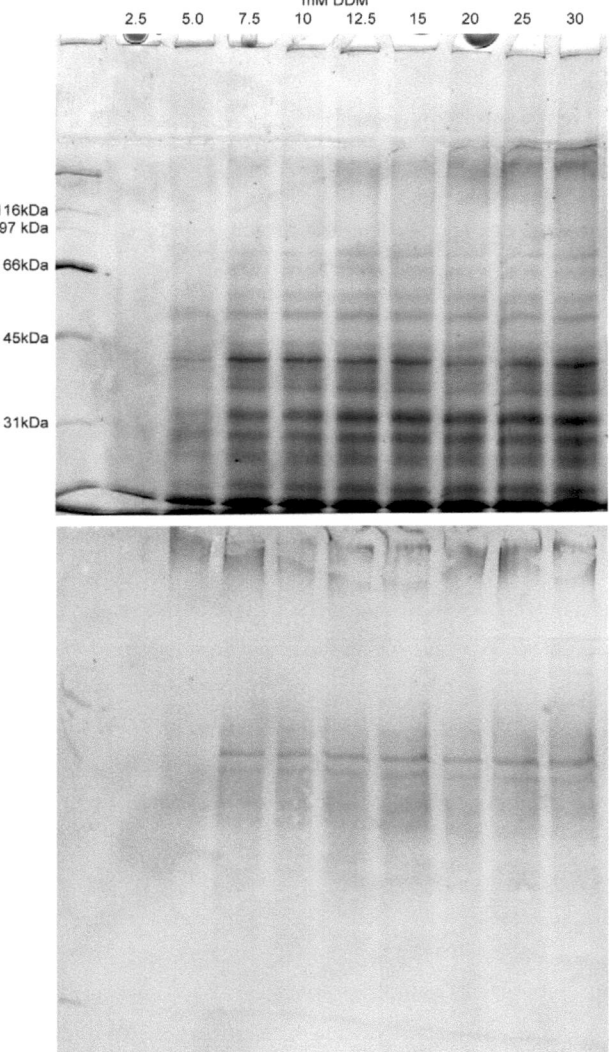

Fig. 2 Silver gel and Western blot of TcHMA4 solubilised with DDM in different concentrations. Lanes from left to right: Marker, 2.5mM, 5mM, 7.5mM, 10mM, 12.5mM, 15mM, 20mM, 25mM, 30mM DDM

In initial tests, isolated TcHMA4 degraded into many small fragments within 1-2 days. Therefore, the effect of different protease inhibitors (phenylmethylsulphonylfluoride = PMSF, PeFa-Block and protease inhibitor cocktails), EDTA and –SH group protecting reductants (DTT, β-mercaptoethanol, Trihydroxypropylphosphine = THP and tris(2-carboxyethyl)phosphine = TCEP) was tested. Protease inhibitors did not lead to any improvement (TcHMA4 degradation bands on the

SDS gel and Western blot). Antioxidants had a strong positive effect, which is understandable in view of the 58 cysteine residues in the TcHMA4 sequence (48 in the 60 kDa main band isolated here, see characterisation below). 10 mM TCEP was by far the best because it led to the smallest amount of degraded protein bands on SDS gels and Western blots.

The final composition of the buffers was:
- isolation buffer: 50 mM HEPES, 250 mM KCl, 10 mM TCEP, pH 6.0.
- solubilisation buffer: 180 mM NaH_2PO_4, 2.0 M NaCl, 10 mM TCEP and 10 mM DDM, pH 6.0.

Purification of TcHMA4

All column materials were tested for the binding of TcHMA4 at pH values of 5 to 9. Among the tested materials, the Ni-IDA column from Macherey-Nagel was the most effective in binding the TcHMA4 protein, leading to a strong band on SDS gel and Western blot in the elutions and hardly any TcHMA4 in the washes. Less binding was observed for the Ni-TED material. Cu-loaded fractogel yielded complete binding but lacked specificity, while Cd- or Zn-loaded fractogel, Talon-resin and Ni-sepharose did not bind TcHMA4 protein at all. After these initial column binding tests, the binding conditions on the column were optimised for pH and salinity, resulting in the use of pH 8.5 and 0.3 M NaCl on the column. Tests for separating TcHMA4 from weakly bound proteins resulted in the use of an exponential gradient of increasing imidazole and decreasing pH for elution.

The resulting protocol was as follows. First, the column was equilibrated with 10 bed volumes of a buffer containing 49.5 mM HEPES, 0.5 mM $NaH_2PO_4H_2O$ (pH 8.5), 300 mM NaCl, 2 mM TCEP and 2 mM DDM. The solubilised crude membrane protein was diluted 1:5 with the same buffer as used for equilibration and the pH was adjusted to pH 8.5. It was loaded onto the Ni-column at a flow rate of 0.1 bed volumes per minute. The unbound proteins were collected in the first wash fraction (designated as W_0). The column was then washed with 20 bed volumes of the same buffer. Subsequently, the bound protein was eluted with equilibration buffer, and an exponential gradient of imidazole increasing from 0 to 2 M and pH decreasing from 8.5 to 6.25. The elution fractions were collected, concentrated with centrifugal concentrators and stored on ice.

The purity of the eluted protein as well as the molecular weight was assessed using SDS-PAGE and Western blotting (Fig.3). TcHMA4 was eluted in high purity at about 320 mM imidazole and pH 6.5. Protein estimation of this purified protein compared to the total amount of protein solubilised from the roots showed that we could isolate about 500 µg of TcHMA4 protein per gram dry weight (about 50 $µg.g^{-1}$ fresh weight) of root material. The purified TcHMA4 represented about 5% of the total protein solubilised from the roots.

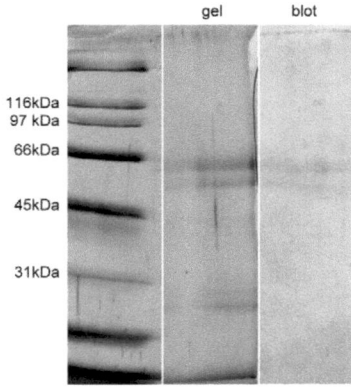

Fig. 3 Purified protein eluted from the Ni-IDA column: silver gel (left) and western blot with TcHMA4 antibody (right).

Characterisation of purified TcHMA4

Gels of crude root extract and of purified protein showed two TcHMA4-specific bands of about 50 and 60 kDa, respectively (Figs. 2,3). In contrast, the mRNA for the *TcHMA4* gene, isolated from *T. caerulescens* plants grown under the same conditions as applied now, predicted a single protein with a size of 128 kDa (Papoyan and Kochian, 2004). This revealed that post-translational processing occurred. The binding of the 60 kDa band to the column and its recognition by the primary antibody (Fig. 3) showed that it contained the C-terminal ca. 550 amino acids of TcHMA4 including the polyhistidine repeat and the region containing 48 cysteine residues, but not the predicted phosphorylation site.

Nevertheless, the activity assay showed that the purified TcHMA4 is active as an ATPase with a turnover rate of 0.85 (µM ATP/µM TcHMA4)/s, verifying the intact state of the protein. This rate could be increased strongly by the addition of Zn^{2+} and Cd^{2+} to the reaction mix (Fig. 4), showing that TcHMA4 is activated by these two metals in concentrations of 0.1 µM. The activation by Zn^{2+} was much stronger compared to Cd^{2+}, indicating that this transporter is mainly responsible for zinc uptake. With increasing concentrations of both metal ions, the activity decreased slightly compared to the value at 0.1 µM, showing substrate inhibition.

The ATPase activity of the isolated protein, combined with the fact that the 50 kDa band was recognised by the primary antibody, therefore indicates that the 50 kDa band is a second subunit that contains the phosphorylation site and the antibody binding site. The finding that sometimes it was lost during purification while the 60 kDa band was always present indicates that the 50 kDa band may not contain the polyhistidine repeat, but is purified along with the 60 kDa band only if the two subunits are bound to each other.

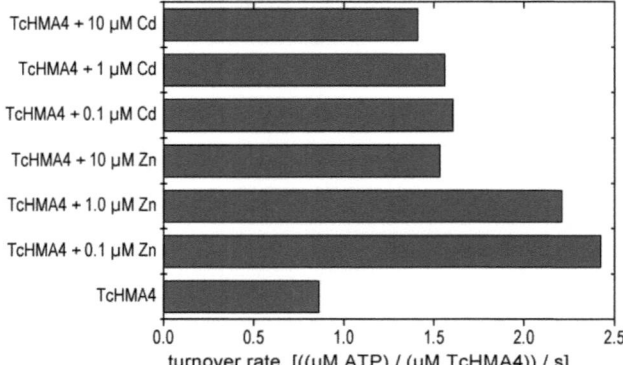

Fig. 4 Activity assay combined with metal activation test

Conclusions

In this project we succeeded to isolate and purify TcHMA4, a Cd/Zn transporting ATPase, from natural overexpression in the roots of *T.caerulescens*. This strategy, in contrast to heterologous overexpression, was chosen to obtain the protein in its native state without problems of misfolding and with all natural post-translational modifications. The latter was important, as the protein, which has a predicted size of 128 kDa based on the cDNA sequence, appears to be post-translationally split, resulting in a main band at ca. 60 kDa and a secondary band at ca. 50 kDa. Purification of TcHMA4 was achieved by systematic testing and optimisation of various parameters involved in solubilisation, column binding and elution. The resulting protocol allowed for isolation and purification of TcHMA4 in high yield, good purity and in its metal inducible active state. Future work will now focus on reconstitution into black membranes and will allow for a more detailed analysis of the transport characteristics of the protein. Spectroscopic and structural investigations are planned to analyse the binding environment of cadmium and zinc and the pumping mechanism.

Acknowledgements

We would like to thank Nina Jagmann for help with establishing the solubilisation conditions. AP is grateful to the Deutscher Akademischer Austauschdienst (DAAD) for a fellowship. MY gratefully acknowledges support by grant 10675035 from the National Natural Science Foundation of China and a fellowship from the Ministry of Education of Baden-Württemberg. HK was supported by the Fonds der Chemischen Industrie (FCI), the Stiftung Umwelt und Wohnen and Degussa-Hüls AG.

2.4. Biochemical and biophysical characterisation yields insights into the mechanism of a Cd/Zn transporting ATPase purified from the hyperaccumulator plant *Thlaspi caerulescens*[†]

Barbara Leitenmaier[1], Annelie Witt[1], Annabell Witzke[1], Anastasia Stemke[1], Wolfram Meyer-Klaucke[2], Peter M.H. Kroneck[1] and Hendrik Küpper[1,3*]

1) Universität Konstanz; Mathematisch-Naturwissenschaftliche Sektion; Fachbereich Biologie; Postfach M665; D-78457 Konstanz; Germany

2) EMBL Outstation Hamburg c/o DESY, Notkestr. 85, D-22603 Hamburg, Germany

3) University of South Bohemia, Faculty of Biological Sciences and Institute of Physical Biology, Branišovská 31, CZ-370 05 České Budejovice, Czech Republic

* to whom correspondence should be addressed. E-Mail: hendrik.kuepper@uni-konstanz.de, Tel.: ++49-7531-884112, Fax: ++49-7531-884533

[†]Financial support for this project was given by the "Landesstiftung Baden-Württemberg", the Stiftung "Umwelt und Wohnen" the "Fonds der chemischen Industrie (FCI)" and the University of Konstanz.

published in Biochimica et Biophysica Acta (section Biomembranes) 1808: 2591-2599

Keywords

metal specificity; hyperaccumulator; natural overexpression; Cd/Zn ATPase; activation energy; temperature optimum

Abbreviations

DDM = n-Dodecyl-β-D-maltoside
E_A = activation energy
EXAFS = extended X-ray absorption fine structure
TCEP = tris(2-carboxyethyl)phosphine
XAS = X-ray absorption spectroscopy

Abstract

TcHMA4 (GenBank no AJ567384), a Cd/Zn transporting ATPase of the P_{1B}-type (=CPx-type) was isolated and purified from roots of the Cd/Zn hyperaccumulator *Thlaspi caerulescens*. Optimisation of the purification protocol, based on binding of the natural C-terminal His-tag of the protein to a Ni-IDA metal affinity column, yielded pure, active TcHMA4 in quantities sufficient for its biochemical and biophysical characterisation with various techniques. TcHMA4 showed activity with $Cu^{(2+)}$, $Zn^{(2+)}$ and $Cd^{(2+)}$ under various concentrations (tested from 30 nM to 10 µM), and all three metal ions activated the ATPase at a concentration of 0.3 µM. Notably, the enzyme worked best at rather high temperatures, with an activity optimum at 42°C. Arrhenius plots yielded interesting differences in activation energy. In the presence of zinc it remained constant (E_A =38 kJ.mol^{-1}) over the whole concentration range while it increased from 17 to 42 kJ.mol^{-1} with rising copper concentration and decreased from 39 to 23 kJ.mol^{-1} with rising cadmium concentration. According to EXAFS the TcHMA4 appeared to bind Cd mainly by thiolate sulfur from cysteine, and not by imidazole nitrogen from histidine.

1. Introduction

Plants that hyperaccumulate heavy metals such as Zn and Cd are currently of great scientific interest as they can be used for cleaning up contaminated soils, a process called phytoremediation (Salt et al., 1995, Raskin et al., 1997). Furthermore, the metal gained in such way can be reused, a fact that is important as the prices for raw metal were rising during the last years and still are. Although many studies have been conducted on the topic already, hardly any metal transporting protein from a plant, hyperaccumulator or not, has been purified in its native state to date. This is surprising as knowledge about these proteins is crucial for understanding the mechanism behind the process of hyperaccumulation. It is known that several families of transporters, like P_{1B}-type ATPases, ZIPs, Nramps, CDFs and CAXs (Hall and Williams, 2003, Williams and Mills, 2005, Küpper and Kroneck, 2005) are involved in the processes of metal uptake into roots, translocation to the leaves and sequestration into large epidermal storage cells. Once in those large storage cells, the metal is finally dumped in the vacuole. There, it can not harm sensitive enzymes as present in mesophyll cells (Küpper et al., 1999, Küpper et al., 2001).

The P_{1B}-type ATPases, a subfamily of the P-type ATPases, are ubiquitous transmembrane proteins that use the hydrolysis of ATP to ADP and phosphate as a driving force to pump metal ions across membranes against an electrochemical gradient. They are found in all kingdoms, including humans (Solioz and Vulpe, 1996). Mutations in these ATPases can lead to Menkes and Wilson disease in humans, which affect the body's ability to maintain the fine balance between copper deficiency and copper toxicity (Bull and Cox, 1994). The fact that several lethal diseases are caused by a malfunction of P_{1B}-type ATPases makes them an even more important research topic.

Plant P-type ATPases consist of eight to twelve transmembrane helices, in which their N and C termini are directed toward the cytoplasm. Between the transmembrane helices a large cytoplasmic loop is located that harbours the phosphorylation site and ATP binding domains (Palmgren and Harper, 1999, Axelsen and Palmgren, 2001).

Although the gene sequences for many plant metal transporters are known, no single crystal structure of a full-length protein and only one biochemical characterisation of a holoenzyme (Parameswaran et al., 2007) exists. Additionally, the copper transporting ATPase ATP7 from humans, has been purified and biochemically characterised (Hung et al., 2007) with the help of heterologous overexpression in insect cells. There are several reasons for the lack of studies on full-length plant metal transporters, one of them is the well-known difficulty of overexpressing membrane proteins homologously without generating misfolding of the protein. Second, in the case of hyperaccumulators, it is difficult to overexpress their metal transporters in another system like *E.coli* as the host would most likely not tolerate high metal concentrations possibly leading to artefacts in the expressed protein as well. Further, in the case of HMAs (named due to their heavy metal associated domain), a very high number of cystein-residues is present in the proteins (Papoyan and Kochian, 2004), making them very sensitive towards oxidation and thus leading to problems with stability, once successfully purified (Parameswaran et al., 2007). In hyperaccumulator plants, many metal transporters are naturally highly overexpressed (Pence et al., 2000, Becher et al., 2004, Hanikenne et al., 2008), making them candidate sources for the extraction of these proteins. HMA is particularly interesting, as it is, according to many recent studies, a key player in the hyperaccumulation phenotype (Verret et al., 2004, Courbot et al., 2007). Additionally, HMA4 from *Thlaspi caerulescens* has 9 consecutive histidine residues located at its C-terminus (Papoyan and Kochian, 2004), offering the possibility to purify this protein based on immobilized metal affinity chromatography.

2. Material and Methods

2.1. Plant material, culture conditions and culture

Thlaspi caerulescens (Ganges ecotype) was grown under conditions that, according to previous investigations (e.g. Parameswaran et al., 2007; Papoyan and Kochian, 2004), lead to high expression levels of metal transporters. All plants were grown in hydroponic nutrient solution with automatic constant media exchange in a controlled environment room, so that the growth conditions could be optimally controlled. Seeds were germinated on a mixture of perlite and vermiculite moistened with deionised water. Three weeks after germination, seedlings were transferred to vessels filled with hydroponic nutrient solution containing 100 µM Zn^{2+} (supplied as $ZnCl_2$). The nutrient solution was aerated continuously using a lab built system and automatically renewed by a programmable peristaltic pump (Ismatec MCP process). Plants were grown with 14 h day length and 22 °C / 18 °C day/night temperature. The photon flux density during the light period followed an approximately sinusoidal cycle with a maximum around 150 µmol.m-2.s-1 and was supplied by full-spectrum discharge lamps.

2.2 Chemicals and Materials

During the optimisation of our original purification protocol (Parameswaran et al., 2007), we realised that contaminations of regular analytical grade chemicals caused problems for further characterisation of TcHMA4. Metal contamination of the used buffer after the dialysis step (see purification protocol below) caused metal loading of the protein, making activity tests or spectroscopic experiments impossible. An unknown contamination in TCEP from other suppliers than chosen by us (see below) caused complete inactivation of the protein and sometimes even degradation visible on gels.

Consequently, all buffer substances for isolation and purification were purchased from Merck, Darmstadt, Germany, in the highest grade available (Suprapur® or Ultrol®), except for the following. Mannitol was purchased from Sigma-Aldrich (St. Louis, MO, USA), in the highest grade available (SigmaUltra). Detergent: n-dodecyl-ß-maltoside (DDM) from Carl Roth, Karlsruhe, Germany and Anatrace (Maumee, OH, USA). Protease inhibitor: Protease Inhibitor Cocktail tablets "complete" EDTA-free (Roche Diagnostics, Mannheim, Germany). Antioxidant: TCEP from Merck (Darmstadt, Germany) or Hampton Research (Aliso Viejo, CA, USA). ATP magnesium salt: also purchased from Merck (Darmstadt, Germany) but with the grade "Low metals Grade". Metal Chelator: Chelex Resin, Bio-Rad (Hercules, CA, USA). Column material: "Protino" Ni-IDA, Macherey-Nagel, (Düren, Germany).

Other materials: Centrifugal protein concentrators: Amicon Ultra-15, membrane with 30 kDa exclusion size (Millipore Corporation, Bedford, MA, USA). Instruments of the following manufacturers were used: Wheat mill: Jupiter 872 with Messerschmidt stainless steel grinding engine (Messerschmidt Hausgeräte GmbH, Königsfeld-Erdmannsweiler, Germany), Ultracentrifuge: Beckmann LE-80 (Beckmann, Palo Alto, CA, USA), AAS: GBC 932 AA (Victoria, Australia).

2.3 Isolation

The isolation of TcHMA4 from roots of *Thlaspi caerulescens* Ganges was carried out as systematically developed (incl. tests of various buffer compositions and protease inhibitors) by Parameswaran et al. 2007, with the following changes: the composition of the isolation buffer has been further optimised to 300 mM Mannitol, 30 mM Hepes and 3 mM $MgCl_2$ as well as "complete" EDTA-free protease inhibitor (1 tablet/ 50 ml buffer), pH 6.0 with KOH. Under liquid nitrogen, the roots were ground to a fine powder only with the wheat mill. We replaced dry ice by liquid nitrogen as the impurities in the dry ice changed from batch to batch and we were not able to get reliable, comparable results. The ratio of root material to isolation buffer was changed to 1:5. In contrast to Parameswaran et al., 2007, we conducted two centrifuge runs of 1 h at 4 °C at 246000*g* instead of only one to remove more soluble proteins. After the second centrifuge run, the supernatant was discarded and the pellet was solubilised for 4-6 h in a, compared to Parameswaran et al., 2007, slightly changed buffer (160 mM NaH_2PO_4, 1.6 M NaCl, 10 mM TCEP, 10 mM DDM (Dodecyl-maltoside), 1 tablet/ 50 ml buffer "complete" EDTA-free protease inhibitor), pH 6.0 with NaOH. The buffer composition has been modified as the now used buffer yielded higher solubilisation efficiency. After the crude extract had been collected, it was transferred into a dialysis tube and dialysed against a buffer containing 300 mM NaCl, 50 mM Hepes, 0.5 mM NaH_2PO_4, 2 mM TCEP, one tablet of "complete" EDTA-free and 10% Chelex, pH 6.0 with NaOH. As the plants were grown in a nutrient solution containing a high concentration of zinc, the dialysis with Chelex was necessary to remove metal in the solution in order to obtain protein not loaded with metal. The mixture was stirred constantly at 4°C for 14 h. This dialysis had to be done to remove metal bound to the protein. The integrity of the protein was checked at all steps during the process via SDS gels and Western blots to make sure that dialysis did not affect its quality. After that, the crude extract was diluted with a buffer as used for dialysis but without Chelex and adjusted to pH 6.0 with NaOH, then centrifuged at the same conditions as after the isolation. With this centrifugation step we were able to remove aggregated protein that would otherwise block the pre-column-filter.

2.4 Purification

As the protein did not efficiently bind to the column at lower pH the supernatant was adjusted to pH 9.0 and loaded onto the Ni-IDA metal affinity column (cooled to 2°C). After that, postwash (buffer composition: 300 mM NaCl, 50 mM Hepes, 0.5 mM NaH_2PO_4, 2 mM TCEP and 0.2 mM DDM plus 2 tablets "complete" EDTA-free, pH 9.0 with NaOH) took place until a stable baseline was reached. Then, TcHMA4 was eluted with a combined gradient of the same buffer as for the postwash at pH 6.0 towards the same buffer with pH 6.0 (adjusted with HCl) including 0.25 M imidazole. The eluted fractions of interest were concentrated for the following characterisation techniques.

2.5 Methods for testing protein identity and purity

Identity and purity of TcHMA4 were controlled via SDS-Page and Western blots as described in Parameswaran et al., 2007, with the following modification: all protein samples were precipitated with 20% TCA for removal of detergent. The pellets were dissolved in sample buffer. After the run, the gels were stained with Lumitein Protein Stain (Biotium, Hayword, CA, USA). Visualisation was achieved using the LumiImager and the LumiCapt Software (Boehringer, Mannheim, Germany). Alternatively, silver staining was used as described in Parameswaran et al., 2007.

2.6 Protein determination

Standards of BSA with known concentration were loaded onto the same SDS gel as samples of TcHMA. Prior to loading, all standards and samples were precipitated with 20% TCA and the pellets were dissolved in the same sample buffer. After the run, the gel was stained with Lumitein Protein Stain (Biotium, Hayword, CA, USA). Quantification was accomplished using the LumiImager and the LumiCapt Software (Boehringer, Mannheim, Germany).

2.7 ATPase activity assay

The activity and activation of TcHMA4 was assayed as described in (Serrano, 1978), with the following important changes in order to obtain more data with a higher reproducibility. In the first step the ATPase was reconstituted into lipid vesicles. For this purpose the ATPase was diluted to 0.18 mg ml^{-1} and then mixed in a ratio 1:5 with phosphatidylcholine from soybean. Reconstitution took place for 20 min at 37°C. For the ATPase reaction, the reconstituted protein was transferred into a buffer containing 0.6 M Tris, 2 M NaCl and 0.1M $MgCl_2$ pH 8.0. The reaction was started with the addition of 5 mM ATP and took place for 40 min at 30°C. A sample without metal addition plus 5 concentrations of copper, zinc and cadmium from 0.1 µM to 3 µM or 10 µM, respectively, were used for measuring activation. Eight phosphate standards from 0 µM to 1200 µM were used for the standard curve. After stopping the reaction with stop solution (0.5% SDS, 0.5% $NH_4Mo(H_2O)_4$, 2% sulfuric acid) , the color reaction was started with the addition of 40 µl/ml 10% ascorbic acid and finally the color development was stopped after exactly 2 min with the addition of

35 µl/ml 34% sodium citrate. Then, the absorbance was measured using a Lambda 750 UV/Vis photometer from Perkin-Elmer (Waltham, MA, USA).

2.7.1 Temperature dependent activity assay

The temperature dependent activity assay was conducted following the same procedure as described above, but only 4 phosphate standards (0 µM, 100 µM, 300 µM and 100 µM) were used and due to the high number of samples, only one metal could be tested per assay. These 10 samples (4 standards, one protein fraction without metal addition and 5 samples of protein with different concentrations of copper, cadmium or zinc) were incubated at 10 different temperatures (generated by a gradient thermostat built for this series of experiments, VLM GmbH, Bielefeld, Germany) ranging from 2°C to 50°C during reconstitution and reaction.

2.8 EXAFS

Sample preparation. For preparing EXAFS samples, micro cuvettes with a volume of approx. 25 µl were used, window size was 12*2 mm. 10 µM (=1.2 ppm) cadmium was added to the unconcentrated protein as Cd-acetate before the concentration process started, yielding an approximate conc. of 15 ppm cadmium in the EXAFS sample. While concentration took place, the non-protein bound metal ran through the membrane, leading to a protein solution that contained <10% (the 1.2 ppm background) free metal and >90% protein-bound metal. After reaching a volume of approx. 50 µl, 10% glycerol was added. With a syringe, the solution was transferred into the cuvettes that were already before sealed with kapton tape, and finally frozen by rapid immersion in supercooled (-140°C) isopentane.

XAS Data Collection. X-ray absorption spectra at the Cd K-edge (26711 eV) were recorded in fluorescence mode on two independently prepared sets of samples in 2007 (dataset A) and 2010 (dataset B). Set A was recorded at EMBL bending magnet beamline D2 of the EMBL Hamburg (c/o DESY, Hamburg, Germany) equipped with a Si(111) double crystal monochromator, a 13 element fluorescence detector (Canberra). Set B was recorded at wiggler station 7-3 of the SSRL (Menlo Park, CA, USA) equipped with a Si(220) double crystal monochromator, a focusing mirror and a 30 element Ge solid-state fluorescence detector (Canberra). For the measurements, the protein samples in their cuvettes (see above) were kept at 20K in a closed cycle cryostat (set A) and at 4 K in a helium cryostat (set B), respectively. Data reduction was performed with KEMP (Korbas et al., 2006) for set A and with EXAFSPAK (http://www-ssrl.slac.stanford.edu/exafspak.html) for set B.

EXAFS Analysis. The extended X-ray absorption fine structures (EXAFS) of both data sets were analyzed using k3-weighting up to k = 10.5 Å-1 with DL-EXCURV (Tomic et al., 2005), which is a freely available version of Excurve (Tomic et al., 2005) under the flagship of CCP3 (www.ccp3.ac.uk). The fit index was used as a measure of the goodness of the fits. *Ab initio* theoretical phase and amplitude functions were generated within EXCURV using Hedin-Lundqvist exchange potentials and van Barth ground states.

2.9 Quantification of metal by AAS

Atomic absorption spectroscopy has been measured on the samples that were previously used for the Cd-EXAFS measurements, i.e. on TcHMA4 that has been loaded with 10 µM cadmium before concentration, yielding protein-bound metal without background.

3. Results

We successfully purified TcHMA in its native state from plant roots. Furthermore, we were able to reconstitute it in artificial lipid vesicles and to conduct activity assays on this reconstituted protein in the presence of cadmium, copper and zinc in various concentrations, elucidating the metal-dependent activation of this protein. Furthermore, we were able to analyse the temperature dependence of this metal-activated TcHMA4 activity. From this temperature dependent activity test, the activation energy of TcHMA4 in the presence of various concentrations of cadmium, zinc and copper was calculated using Arrhenius plots. Furthermore, the binding sites of cadmium in TcHMA4 were analysed by EXAFS spectroscopy, and via AAS of these samples we could determine the metal to protein ratio.

3.1 Purification - difficulties with instability of TcHMA4

The elution from the metal affinity column yielded, when all problems described below were kept at a minimum, a rather pure preparation of active TcHMA, as identified by SDS gel and western blot (Fig. 1). The yield of this purification was approximately 1.2 mg TcHMA4 per 60 g of root material.

Despite this success, the extreme instability of the protein, however, was a major problem throughout the purification. Approximately 12 h after purification, the first activity test was carried out and after an additional 6 h more, the protein had lost more than half of its activity. On the next day, 36 h after purification, it was completely inactive. This meant first that for each repeat, fresh protein had to be purified leading to unavoidable variations between the samples. As explained in detail in Parameswaran et al. 2007, we tested various reducing agents for increasing the stability of TcHMA4, with TCEP yielding the best overall result. In the presence of other reducing agents, TcHMA4 was even less stable or did not show any activity at all, or they interfered with the activity assay (main reason not to use DTT). We tested various batches of TCEP from different suppliers finding that the purity of TCEP was crucial to obtain high activity. Using slightly less pure batches of TCEP, the ATPase activity was completely inhibited, only TCEP with a purity higher than 99.5% from Merck or Hampton Research could be used for our tests.

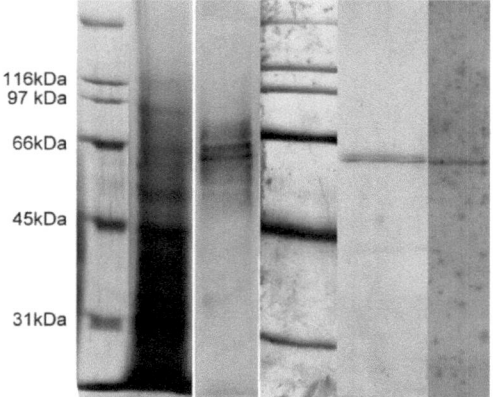

Fig. 1 Identity and purity of TcHMA4 tested by SDS gel and western blot with specific antibody. Left: molecular weight marker, silver stained SDS gel of crude extract, western blot of crude extract with a characteristic band at 60kD. Right: molecular weight marker, silver stained SDS gel of purified TcHMA4, western blot of purified TcHMA4.

3.2 ATPase activity assay with different metal concentrations

The ATPase activity of TcHMA4 could be increased (compared to the "as isolated" state without addition of metals, displayed as "0.01 µM" in the graphs) by both hyperaccumulated metals of *T. caerulescens*, cadmium and zinc, but also by copper (see Fig. 2). Nevertheless, the concentration range that yielded highest activity differed strongly for the different metals. In the case of cadmium and zinc, the highest increase in activity was observed in the presence of 0.03 µM metal and the level of activation did not change much until a concentration of 1 µM Cd^{2+}, or 0.3 µM Zn^{2+} respectively, was added. A maximal turover rate of 300 s^{-1} was reached with fully activated TcHMA4, with zinc as well as with cadmium. Higher concentrations of Cd or Zn did not further increase activity. In the case of cadmium, starting from 3 µM, the activity decreased strongly, in the case of zinc, the decrease started at 1 µM already. For copper, a clear maximum of activity was observed in the presence of 0.1 µM Cu^{2+}, with higher concentrations the ATPase activity decreased. The addition of 10 µM copper decreased the ATPase activity even below the level without additional metal in the sample (estimated residual metal content 0.01 µM). For cadmium and zinc, the decrease of activity was less severe.

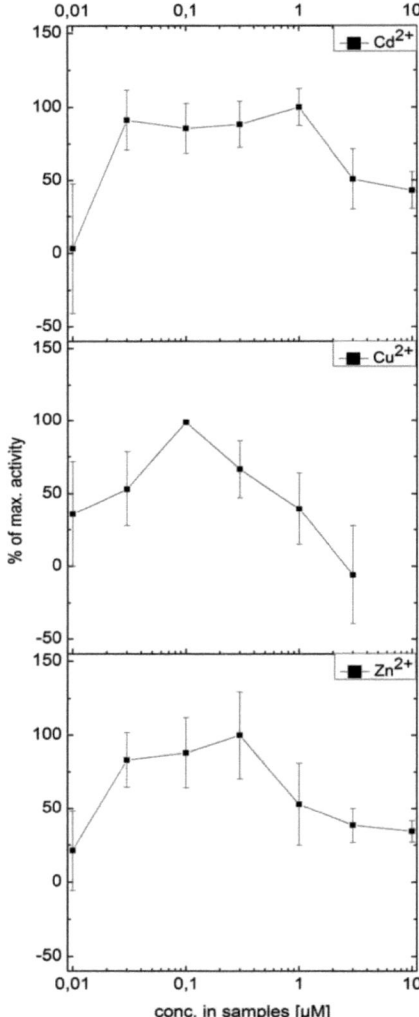

Fig. 2 Colorimetric phosphate determination for measuring activation of TcHMA4 in % of max. activity for different metals in various concentrations. Reaction time 40 minutes at 30°C.

3.3 ATPase activity assay with different metal concentrations and a temperature gradient

It was very difficult to perform this kind of experiment, because with increasing temperatures the ATP that had been added to start the ATPase reaction hydrolyzed more easily. To circumvent this problem, we used an individual phosphate calibration for each temperature. For all three metals, we found an increased activity towards higher temperatures. For cadmium and copper, the highest activity was reached at almost 50°C with concentrations of 0.03 µM to 0.1 µM, while for zinc the highest activity was reached at approximately 50°C and a concentration of 0.5 µM (see Fig. 3). A clear activation in the zinc concentration range from 0.1 µM to 1 µM can be observed, this phenomenon is especially clearly visible at temperatures from 35°C upwards.

When normalising the obtained data for zinc on the value without metal added for highlighting the influence of metal concentrations, it becomes obvious that the activation (activity with metal divided by activity without metal) is constant over a large range of temperatures, only at very high temperatures (>45 °C) it drops. At very low temperatures (<10°C), the inhibition at high metal concentrations becomes less visible (see Fig. 3).

Calculating the activation amplitudes, i.e. the activity with metal minus the activity without metal added, a temperature of 42°C and a zinc concentration in the range of 0.03 µM to 3 µM yielded highest activation compared to the activity achieved without metal added (Fig. 3), no major change in the temperature optimum has been observed for the three tested metals, cadmium, zinc and copper (not shown).

Using the obtained absolute phosphate concentrations in the presence of different metal concentrations, Arrhenius plots were calculated (see Fig. 4). These plots allowed us to determine the activation energy of TcHMA4. In the presence of various concentrations of zinc, the activation energy (E_A = 38 kJ.mol^{-1}) remained constant over the whole concentration range. In contrast, with very low concentrations of cadmium, the activation energy was highest (E_A = 39 kJ.mol^{-1}), decreasing towards higher concentrations with the lowest activation energy (E_A = 23 kJ.mol^{-1}) required at 3 µM cadmium. For copper, the activation energy strongly increased with increasing metal concentration, from (E_A = 17 kJ.mol^{-1}) at 0.03 µM Cu^{2+} to (E_A = 42 kJ.mol^{-1}) at 3 µM Cu^{2+}.

3.4 EXAFS measurements

Two independent cadmium EXAFS measurements on TcHMA4 have been carried out, one in 2007 at the DESY in Hamburg, Germany, and a second one in 2010 at the SSRL, Stanford, CA, USA. Both spectra obtained show a very similar ligand environment for cadmium: a characteristic peak at 2.5 Å shows that by far the largest contributor is sulfur (see Fig. 5). The refinements with both the numbers and distances of ligands as free parameters yielded four to five sulfur ligands at a distance of 2.5 Å from the central Cd^{2+}. No significant contributions from nitrogen/oxygen ligands were detected in the first ligand shell, and no contributions of the characteristic multiple scattering features of the imidazole ring of histidine (or the imidazole in the elution buffer) were detected in the spectra. For a detailed summary of the refinement parameters see table 1. These results show that the greatest part of the cadmium in TcHMA4 is bound by cysteines and not, as partly expected, by the natural histidine-rich C-terminus

When protein for an EXAFS sample was purified, almost the whole batch was needed for one sample and we were not able to do comprehensive additional tests (e.g. the temperature-dependent activation) on this exact batch of protein. We could not determine the metal concentration present in the EXAFS samples before actually using them for EXAFS, because metal determination via AAS or ICP-MS would have consumed a major part of the sample itself. Doing this analysis after the EXAFS measurements, we found 20±16 cadmium ions/mol TcHMA4 after loading (equilibration) of the protein with 10 μM cadmium. The high error was not mainly caused by different efficiencies of metal loading between preparations but to the most part by detergent interference with the quantification of the protein in attempts of direct quantification on non-precipitated protein, and problems with sample losses and incomplete re-solubilisation when removing the detergent via precipitation.

Table 1
Results of the refinement of the EXAFS spectra using the DL-Excurve program, with the number of ligands, their distances and the Fermi energy as free parameters. N and O are used indistinctively, since their scattering phases are similar. The graphs of the fits are shown in Fig. 5.

EF = Fermi energy shift with respect to E_0=eV used in the background subtraction, defines the threshold for the EXAFS spectra (Rehr and Albers, 2000). This value was refined for every sample.
FI = Fit index with k^3 weighting.
se = Mathematical standard errors of the refinement (two sigma level). The error of the EXAFS approach as such is higher, this is revealed by the differences between samples of the same type.
*: Values were constrained to be identical for these contributions. Deviations from this simplification do not improve the model significantly.

Sample	Number/Type of ligands (±se)	Distance (± se) [Å]	$2\sigma_i^2$ (± se) [Å²]	EF (± se)	FI
Set A (2007)	0.1 (±0.2) N/O	2.05 (±0.39)	0.02 (fixed)	-2.3 (±1.6)	1.0
	4.7 (±0.3) S	2.51 (±0.01)	0.02 (fixed)		
Set B (2010)	0.2 (±0.8) N/O	2.08 (±0.77)	0.02 (fixed)	1.5 (±3.9)	4.8
	3.7 (±0.9) S	2.47 (±0.05)	0.02 (fixed)		

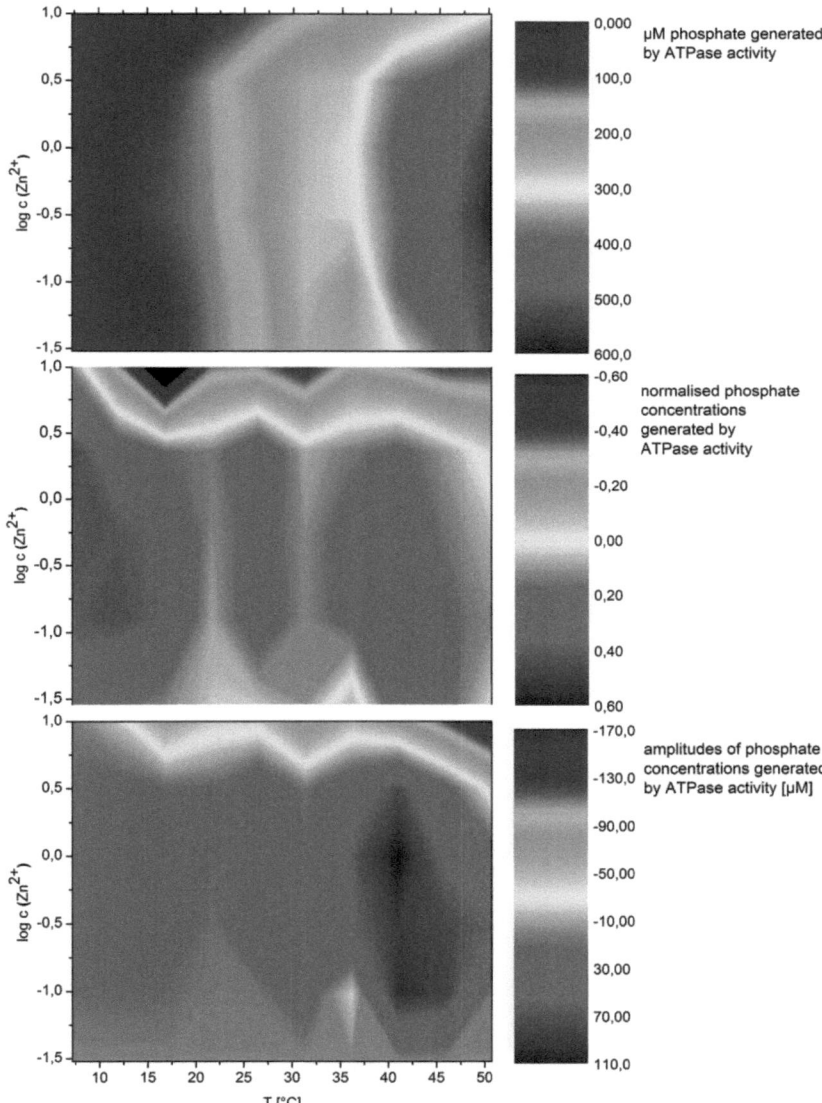

Fig. 3 Temperature and zinc dependent activity of TcHMA4, measured via colorimetric determination of phosphate after a reaction time of 40 minutes. *Top:* TcHMA4 activity. *Middle:* activation ratio, calculated by normalising the activity of each point to the activity of the basic ("as isolated", i.e. without metal addition) activity of the respective temperature. *Bottom:* activation amplitudes, calculated by subtracting the basic activity of the respective temperature activity from each point of the activities with metal addition.

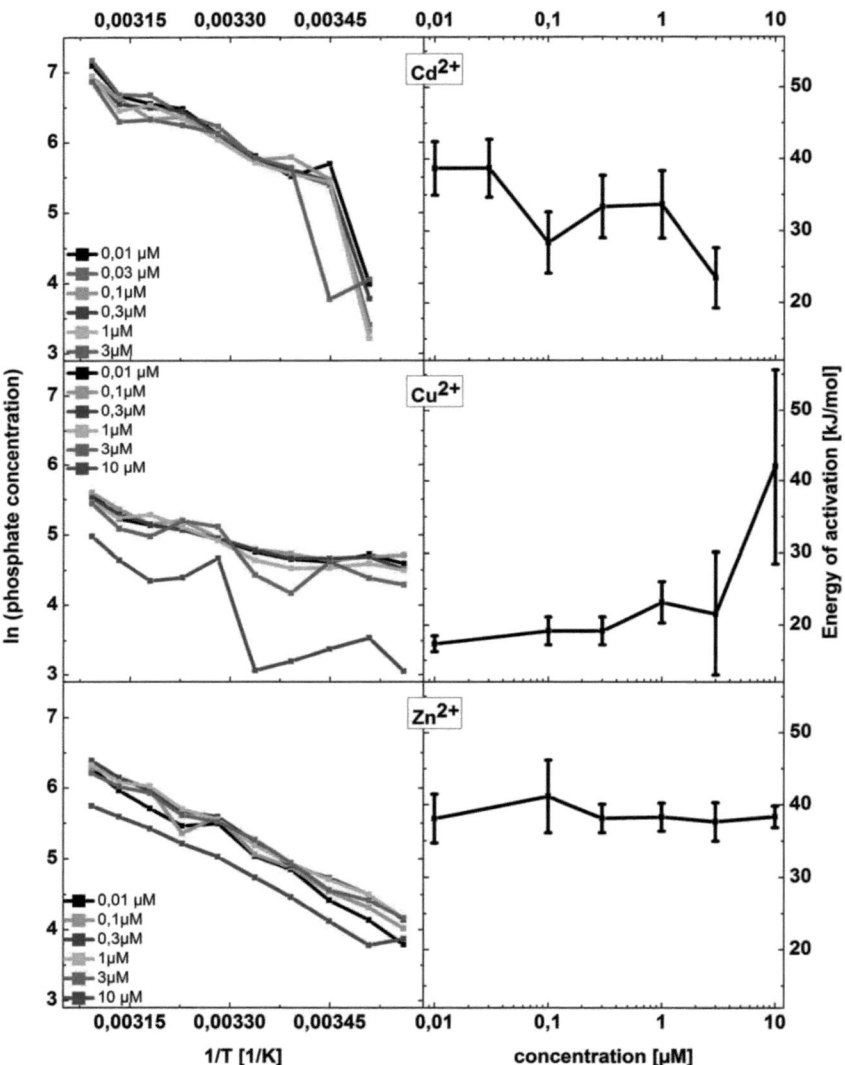

Fig. 4 Arrhenius plots (left) and energies of activation (right) derived from temperature and metal dependent activity assays for Cd, Cu and Zn.

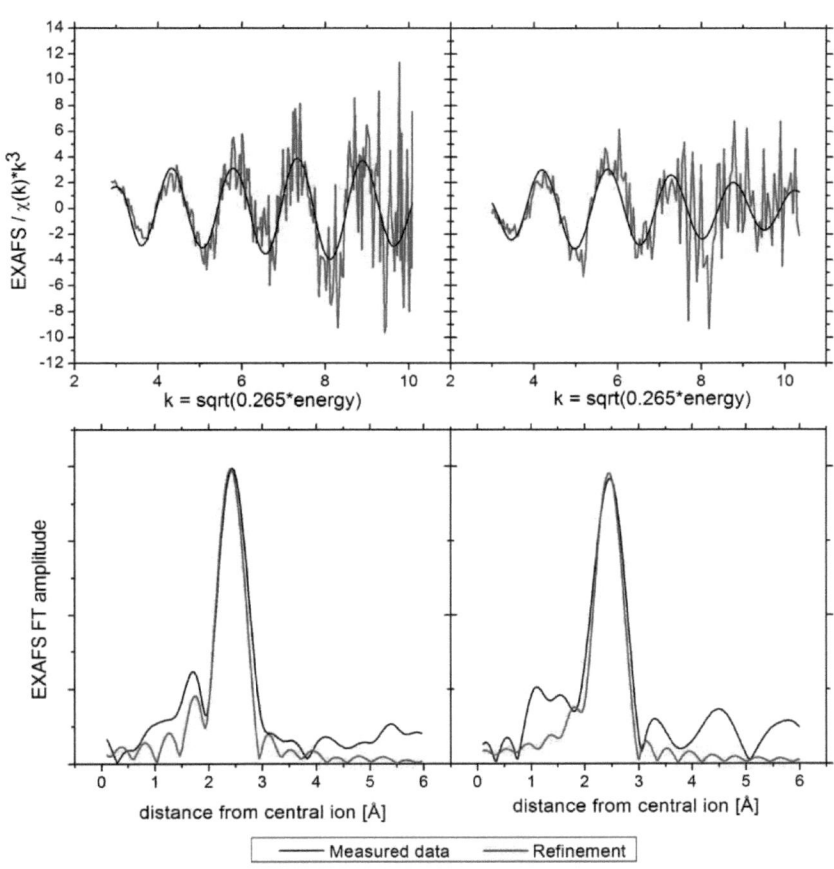

Fig. 5 EXAFS measurements on TcHMA4: raw data with refinements (upper) and Fourier transformed datasets (lower).

4. Discussion

While previous studies on plant metal transport proteins dealt with expression or metal transport on the cell/tissue level, the isolation and purification of TcHMA4 allowed us to investigate important characteristics of such a transporter in a controlled chemical environment. This is the first report of these characteristics for a plant heavy metal (P_{1B}-type) ATPase.

4.1 ATPase activity assay with different metal concentrations

TcHMA4 showed its strongest activation with 0.1 μM copper, in the case of cadmium with concentrations between 0.03 μM and 1 μM, for zinc the optimal range started at 0.03 μM and activity decreased with 1 μM and higher concentrations. As zinc and cadmium are hyperaccumulated by *T. caerulescens* up to a concentration of 3% in the above-ground tissues for zinc and 1% for cadmium (Mijovilovich et al., 2009), while copper is not accumulated, it is interesting to see that the concentrations needed for maximal activity do not differ more. It is assumed that TcHMA4 is involved in the loading of cadmium and zinc into the xylem (Papoyan and Kochian, 2004), through which the metals are transported into the shoot and into their final storage site, the vacuoles of large epidermal cells (Küpper, 1999). Apart from a relatively small number of resistant individuals (Roosens et al., 2004), *T. caerulescens* does not tolerate excess copper concentrations in the soil and clearly shows symptoms of toxicity at copper concentrations starting below 10μM. The activation of TcHMA4 by copper implies that, if present in the soil, TcHMA4 would load copper into the xylem as it does for Cd and Zn. There or at a later point in the shoot the excess of copper that is not used for various copper depending enzymes has to be complexed and transported back to the root, where efflux must take place in order to prevent severe damage mainly to the photosynthetic apparatus. In this complexation of copper, methallothioneins might play an important role (Roosens et al., 2005, Mana-Capelli et al., 2003).

Hung et al. showed in 2007 in similar experiments with an human copper ATPase, MNK or ATP7a, that a concentration of 1 μM copper was required for full activation of the protein, while in the presence of 1 μM zinc the protein showed only 60% of its activity, with 1 μM cadmium the activity even decreased to 10% of its maximum activity (Hung et al., 2007). Mana-Capelli et al. (2003) successfully purified the Cu-ATPase CopB from *Archaeoglobus fulgidus* and showed a maximum of activity of the enzyme in the presence of 1 μM Cu^{2+}, while the activity strongly decreased with the addition of Cu^+ and even more so with cadmium and zinc (Banci et al., 2002). As to date only structures of subdomains of P_{1B}-type ATPases are available (Sazinsky et al., 2006, Banci et al., 2008) but complete structures of any holoproteins are lacking, it remains unknown how these ATPases are able to clearly distinguish between divalent metal cations and in the case of CopB even between the same metal in different oxidation states.

In our experiments, at a concentration of 3 µM copper, TcHMA4 showed less activity than in the sample without additional copper. The phosphate concentration in this case dropped slightly below 0, suggesting that either the protein denatured taking up phosphate, or (more likely) switched from ATPase acitivity to ATPsynthase activity as it is known from other ATPases including the Menkes copper-transporting P_{1B}-type ATPase (Hung et al., 2007).

4.2 ATPase activity assay with different metal concentrations and a temperature gradient

TcHMA4 shows its highest activity at >40°C, with maximum activation at 42 °C. A similar temperature optimum of 40°C has been reported for an K^+ATPase from *Beta vulgaris* L. (red beet) storage roots (Briskin and Poole, 1983). Also at temperatures below 30°C TcHMA4 still pumps metal, at a lower rate than at 42°C, but with the same ratio of metal-dependent activation as at the 42°C optimum. This makes sense because *T. caerulescens* Ganges is a perennial plant (Jiménez-Ambriz et al., 2007) and originates from Southern France, so that its enzymes need to tolerate low temperatures in winter and rather high temperatures in summer. Activity of TcHMA4 over a temperature range from 7.5°C to 50°C is a prerequisite for tolerating elevated Cd/Zn concentrations throughout the year.

The data obtained from activity tests at different temperatures were further used for Arrhenius plots, yielding interesting insights into the activation energy (E_A) required by TcHMA4 for pumping metal over the cytoplasmic membrane. In the presence of cadmium the activation energy required for the function of TcHMA4 decreased towards higher metal concentrations and thus, higher concentrations must facilitate the transport of cadmium over the plasmamembrane. This corresponds with the activation curves shown in Fig. 2, where, starting with a concentration of 0.03 µM, cadmium strongly activated the enzyme compared to the sample without additional metal. At a concentration of 3 µM cadmium, the activity was only 50% of the maximum activity, although the activation energy required for function at high concentrations is lower. It is possible that at high concentrations of cadmium the ATPase starts to denature, and thus loses its original affinity properties yielding unspecific pumping. For copper, comparably low activation energies (below and around 20 kJ.mol^{-1}) were observed over a broad range of concentrations. A strong increase in E_A to 42 kJ.mol^{-1} becomes visible at 10 µM and although the standard error for this value is rather high, the increase is still significant. For zinc, the activation energy stays rather constant at a value just below 40 kJ.mol^{-1}, no dependency on the concentration of zinc could be observed. In fact, the 10 µM used as highest concentration in our TcHMA4 activity assay is not at all toxic for *T. caerulescens*, where Zn toxicity only starts in the mM range. Comparison of E_A values obtained in other studies is difficult as until now, no E_A values for P_{1B}-type ATPases from any plant have been reported. Generally, biochemical characterisations of holoproteins from this family are lacking. Even for H^+-ATPases and Ca^{2+}-ATPases, data are rare. There are several studies reporting activation energy values from fish proteins, ranging from 15 kJ.mol^{-1} to 35 kJ.mol^{-1} (Landeira-Fernandez et al., 2004) and one study observing the temperature-dependent activity of an H^+-ATPase from protoplasts of *Commelina* ("dayflower") where no E_A value has been determined, but an Arrhenius plot has been shown from which one can calculate an E_A value of 8 kJ.mol^{-1} (Willmer et al., 1995). Furthermore, one publication describes E_A values between 30 kJ.mol^{-1} to 80 kJ.mol^{-1} for an ATPase from sugar beet roots in the presence of various concentrations of Na^+ and Mg^{2+} (Lindberg et al. 1976).

4.3 EXAFS measurements

Both the C-terminal histidine stretch and the many cysteines in the structure of TcHMA4 have been proposed as potential metal binding sites (Papoyan and Kochian, 2004, Courbot et al., 2007, Argüello et al., 2007)). To test these hypotheses, EXAFS measurements on Cd-loaded TcHMA4 were performed. It was very difficult to get a sufficient amount of protein for these measurements. As the main reason, due to the detergent present in the samples, it was not possible to concentrate the samples higher without disturbing the protein integrity (tested by activity tests). This was a problem because for EXAFS a rather high metal concentration in the sample is required. The metal concentration of 15 ppm in our samples is far below the ideal concentration of 1 mM, a fact that certainly influences the signal to noise ratio, but is still in a range for proper signal detection. The problem with low concentrations of metals in protein and plant samples is a common one but has been circumvented in a few interesting cases (Roosens et al., 2004, Lane et al., 2005) by increasing the measuring time. Low concentrated samples have to be measured for a comparably long time (often up to several days), making a potent cooling system a necessity. The number of beamlines available for such samples is very small, which might be a reason why only a small number of comparable studies have been carried out so far.

For cadmium, we have measured and analysed two independent samples from two preparations of TcHMA4 originating from different batches of root material, the two samples are referred to batch a and batch b in the following. The Fourier Transforms of both datasets show a very clear peak at 2.5 Å, characteristic for sulfur ligation. This suggests that the dominating fraction of cadmium in the protein samples must be bound to cysteins and not to histidines, as hardly any contribution of nitrogen is visible in the Fourier transformed data sets. As the metal had been added in excess and during the concentration all unbound cadmium had been washed out, it is not likely that the natural C-terminal histidine stretch (that we used for purification) binds cadmium with any significant affinity. Currently, it is not possible to determine the particular cysteins involved in cadmium binding, providing a very important question for future research as accumulations of cysteines are very abundant in the structure of TcHMA4 and most likely not all of these cysteine clusters have the same affinity for cadmium.

5. Conclusions

We successfully investigated the activity of the TcHMA4 holoprotein purified from hyperaccumulator roots, yielding maximal activity of the ATPase in presence of 0.1 µM copper and a wider concentration range in the case of zinc and cadmium. Based on Arrhenius plots we showed that the energy of activation rises for higher copper concentrations while it stays constant for the investigated zinc concentrations. Temperature dependent activity assays showed that the enzyme is not only more active at higher temperatures but that the metal concentration is still important for activity (shown for zinc). This suggests that TcHMA4 does not become unspecifically active at high temperatures but is still able to prevent uncontrolled ATP consumption. EXAFS measurements proved that cadmium in TcHMA4 is mainly bound to cysteins and not to histidines, suggesting that the C-terminal his residues could function as a trap for cadmium while cysteins must be involved in transporting the metal through the membrane. With the method established for purifying the protein in its active state it will now be possible to visualise cadmium and eventually also zinc transport via TcHMA4 over a membrane using specific fluorescent dyes.

6. Acknowledgements

We would like to thank Nils Schoelzel and Michael Mayer for excellent plant care and Timo Witt, Birte Baudis and Alexander Schönborn for help during some of the experiments. Wolfram Welte is gratefully acknowledged for useful comments and discussions, especially concerning different solubilising strategies. Portions of this research were carried out at the Stanford Synchrotron Radiation Lightsource, a national user facility operated by StanfordUniversityon behalf of the U.S. Department of Energy, Office of Basic Energy Sciences. The SSRL Structural Molecular Biology Program is supported by the Department of Energy, Office of Biological and Environmental Research, and by the National Institutes of Health, NationalCenterfor Research Resources, Biomedical Technology. The authors are grateful to SSRL for providing beamtime and thank Erik Nelson and colleagues for their excellent support during the beamtime.

2.5. Zn EXAFS and UV/Vis spectroscopy of TcHMA4 – preliminary results

Barbara Leitenmaier[1], Annelie Witt[1], Annabell Witzke[1], Anastasia Stemke[1], Wolfram Meyer-Klaucke[2], Peter M.H. Kroneck[1] and Hendrik Küpper[1,3]

1) Universität Konstanz; Mathematisch-Naturwissenschaftliche Sektion; Fachbereich Biologie; Postfach M665; D-78457 Konstanz; Germany

2) EMBL Outstation Hamburg c/o DESY, Notkestr. 85, D-22603 Hamburg, Germany

3) University of South Bohemia, Faculty of Biological Sciences and Institute of Physical Biology, Branišovská 31, CZ-370 05 České Budejovice, Czech Republic

Some interesting results regarding the Cd/Zn ATPase TcHMA4 have not been reproduced so far and are, therefore, not in a publishable state yet. Therefore they are not included in any publications or manuscripts ready for submission. Still, they are interesting and very promising giving ideas how to continue the project in the future.

2.5.1.

UV/Vis titration in the non-loaded state ("as isolated") and after loading with cadmium
For materials and methods see 5.1.

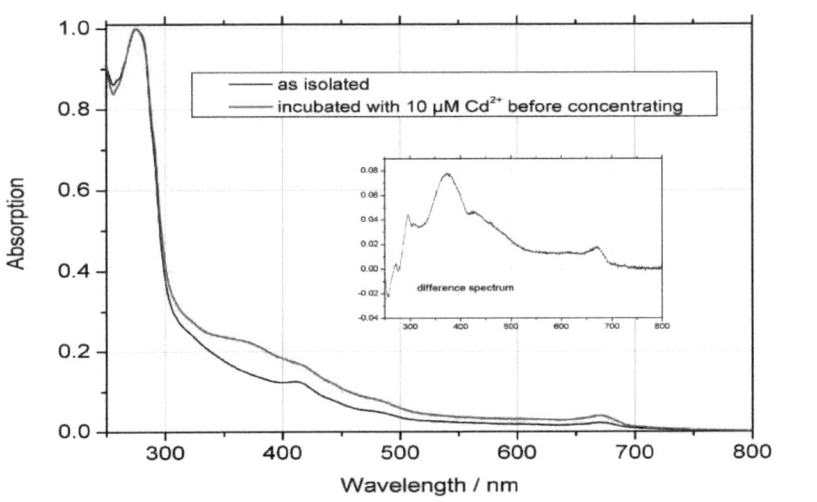

After creating the difference spectrum, clearly hints for charge transfers are visible. After titration with increasing cadmium concentration, saturation of the protein with metal can be calculated, once more data are present. The problem here is that extremely pure protein is necessary, as otherwise the metal would bind to all other proteins in the mixture.

2.5.2.

Additionally to the EXAFS measurements with cadmium, one of these samples has been measured for zinc as well. Surprisingly, zinc was detectable showing that the protein must have been naturally preloaded, as only cadmium has been added after dialysis against chelex and subsequent purification. Still, the zinc EXAFS shows a very interesting pattern:

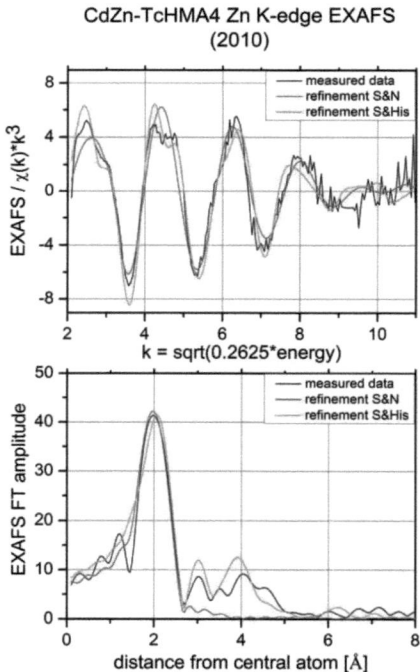

In contrast to the fourier transformed cadmium data, the zinc data do not show the peak at 2.5 Å suggesting sulfur as a ligand. Here, a clear peak at 2 Å is visible, resulting from nitrogen. This nitrogen peak strongly suggests histidine as ligand of zinc in TcHMA4. As the protein has a natural histidine stretch at the C-terminus, this finding makes sense and can explain the slightly different affinities of TcHMA4 towards cadmium and zinc.

Imidazole was present in the sample as it was used in the gradients for purification of protein for EXAFS samples. Its contribution to the signal is negligible, as zero controls with the buffer only, including the same concentration of imidazole and non-protein-bound zinc as the samples, did not yield any recordable EXAFS signal.

2.6. Complexation and toxicity of copper in higher plants (II): Different mechanisms for Cu vs. Cd detoxification in the Cu-sensitive Cd/Zn hyperaccumulator *Thlaspi caerulescens* (Ganges ecotype)

Ana Mijovilovich[1], Barbara Leitenmaier[2], Wolfram Meyer-Klaucke[3], Peter M.H. Kroneck[2], Birgit Götz[2], and Hendrik Küpper[2, 4]

1) University of Utrecht, Department of Inorganic Chemistry and Catalysis, Sorbonnelaan 16, 3584 CA Utrecht, The Netherlands

2) Universität Konstanz; Mathematisch-Naturwissenschaftliche Sektion; Fachbereich Biologie; Postfach M665; D-78457 Konstanz; Germany

3) EMBL Outstation Hamburg c/o DESY, Notkestr. 85, D-22603 Hamburg, Germany

4) University of South Bohemia, Faculty of Biological Sciences and Institute of Physical Biology, Branišovská 31, CZ-370 05 České Budejovice, Czech Republic

published in Plant Physiology 151: 715-731

ABSTRACT

The Cd/Zn hyperaccumulator *Thlaspi caerulescens* is sensitive towards copper toxicity, which is a problem for phytoremediation of soils with mixed contamination. Copper levels in *T. caerulescens* grown with 10 μM Cu^{2+} remained in the non-accumulator range (<50ppm), and most individuals were as sensitive towards copper as the related non-accumulator *T. fendleri*. Obviously, hyperaccumulation and metal resistance is highly metal-specific. Cu-induced inhibition of photosynthesis followed the "sun reaction" type of damage, with inhibition of the PSIIRC charge separation and the water splitting complex. A few individuals of *T. caerulescens* were more copper resistant. Compared to Cu-sensitive individuals they recovered faster from inhibition, at least partially by enhanced repair of chlorophyll-protein complexes but not by exclusion since the content of copper in their shoot was increased by about 25%. Extended X-ray absorption fine structure (EXAFS) measurements on frozen-hydrated leaf samples revealed that a large proportion of copper in *T. caerulescens* is bound by sulfur ligands. This is in contrast to the known binding environment of cadmium and zinc in the same species, which is dominated by oxygen ligands. Clearly, hyperaccumulators detoxify hyperaccumulated metals differently compared to non-accumulated metals. Furthermore, strong features in the Cu-EXAFS spectra ascribed to metal-metal contributions were found, in particular in the Cu-resistant specimens. Some of these features may be due to Cu-binding to metallothioneins, but a larger proportion seems to result from biomineralisation, most likely Cu(II)-oxalate and Cu(II)-oxides. Additional contributions in the EXAFS spectra indicate complexation of Cu(II) by the non-proteogenic amino acid nicotianamine, which has a very high affinity for Cu(II) as further characterised here. EXAFS features characteristic for Cu-phytochelatins were not detected.

Key words: Brassicaceae, copper, heavy metal speciation, metal-induced stress, phytoremediation, phytomining, X-ray absorption spectroscopy

Abbreviations:

Antenna connectivity = the likelihood of energy transfer between antennae of different photosystems (PS II and/or PS I)

CA = component analysis. In this study we use this term for the fitting of EXAFS spectra with a linear combination of the EXAFS spectra of model compounds.

DW = Debye-Waller factor of the EXAFS refinements, it accounts for both structural and thermal disorder of the metal ions coordination shells (Binsted 1992)

EF = Fermi energy of the EXAFS refinements, defines the threshold for the EXAFS spectra (Rehr and Albers, 2000).

EPR = Electron Paramagnetic Resonance

EXAFS = extended X-ray absorption fine structure

FKM = Fluorescence Kinetic Microscope

FI = Fit index of the EXAFS refinements = sum of the square of the residuals - this is what is minimised in the refinement.

F_0 = minimal fluorescence yield of a dark adapted sample, fluorescence in non-actinic measuring light

F_m = maximum fluorescence yield of a dark-adapted sample after supersaturating irradiation pulse

F_m' = maximum fluorescence yield of a light-adapted sample after supersaturating irradiation pulse

F_p = fluorescence yield at the P level of the induction curve after the onset of actinic light exposure

F_v = variable fluorescence; $F_v = F_m - F_0$

$F_v/F_m = (F_m-F_0)/F_m$ = maximal dark-adapted quantum yield of PSII photochemistry

$\Phi_{PSII} = \Phi e = (F_m'-F_t')/F_m'$ = the light-acclimated efficiency of PS II (Genty et al. 1989). In the current manuscript the use of this parameter is extended to the relaxation period after the end of actinic light to analyse the return of the system to its dark-acclimated state as measured by F_v/F_m.

k = wave number of the photoelectron, proportional to the square of the energy difference from the threshold energy (Efermi).

k-range = energy range above the X-ray absorption edge of the metal where EXAFS was analysed.

LHC = light harvesting complex

Light saturation: measured by the increased amplitude of F_p relative to F_m after subtraction of F_0. $(F_p-F_0)/(F_m-F_0)$ is mostly dependent on the ratio of functional antenna molecules to functional reaction centres and electron transport chains. Under constant actinic irradiance for measuring F_p, a large antenna capturing photons and delivering them to its reaction centre will cause more of the "electron traffic jam" that leads to F_p than a small antenna.

multiple scattering = contributions in EXAFS that originate not from direct interaction between the central ion and another atom, but from interactions between the different atoms of a ligand molecule. This can only be observed in very rigid (parts of) ligand molecules, e.g. the imidazole ring of histidine. Further, the contribution is most significant for forward scattering at angles close to 180 degrees, which is the case for the atoms in an imidazole ring..

NA = nicotianamine

NPQ = non-photochemical quenching, in this manuscript used as an acronym for the name of this phenomenon. In this manuscript, we measure non-photochemical quenching as $q_{CN} = (F_m - F_m')/F_m =$ "complete non-photochemical quenching of Chl fluorescence", i.e. with normalisation to F_m.

Pheo = pheophytin

PSII = photosystem II

RC = photosynthetic reaction centre

XAS = X-ray absorption spectroscopy

XANES = X-ray absorption near edge structure

Z = atomic number

INTRODUCTION

Many heavy metals are well known to be essential microelements for plants, but elevated concentrations of these metals cause toxicity (reviewed e.g. by Prasad and Hagemeyer, 1999, Küpper and Kroneck, 2005), as explained in more detail in our companion paper on copper complexation and toxicity in *Crassula helmsii*. Plants developed a number of strategies to resist this toxicity, including active efflux, sequestration, and binding of heavy metals inside the cells by strong ligands. Among the Zn and Cd-hyperaccumulators (Brooks, 1998; Lombi et al. 2000) the best known species is *Thlaspi caerulescens* J.&C. Presl., which has been proposed as a hyperaccumulator model species by several authors (Assunção et al. 2003; Peer et al. 2003, 2006). An enhanced uptake of metals into the root symplasm was found in *T. caerulescens* compared to the related non-accumulator *T. arvense* (Lasat et al. 1996, 1998), and a reduced sequestration into the root vacuoles was associated with the higher root to shoot translocation efficiency of *T. caerulescens* (Shen et al. 1997; Lasat et al. 1998). This is likely related to elevated expression of xylem loading transporters in the roots (e.g. Papoyan and Kochian, 2004; Weber et al. 2004). One of these, the Cd/Zn pumping P_{1b} type ATPase TcHMA4 was recently purified as a protein, which revealed post-translational processing and its biochemical characterisation showed Cd and Zn transport affinity in the submicromolar range (Parameswaran, Leitenmaier et al. 2007). Studies of cellular metal compartmentation have shown that in most hyperaccumulators the metal is sequestered preferentially into compartments where it does no harm to the metabolism. These are, in most cases studied so far, the epidermal vacuoles (Küpper et al. 1999, 2001; Frey et al. 2000, Bidwell et al. 2004; Broadhurst et al. 2004), where concentrations of several hundred $mmol.l^{-1}$ can be reached in the large metal storage cells (Küpper et al. 1999, 2001). The latter showed that hyperaccumulation must be mediated by active pumping of the heavy metals into their storage sites, which was shown to be achieved by an extremely increased expression of metal transport proteins in leaves of hyperaccumulators compared to non-accumulators (Pence et al. 2000, Assunção et al. 2001; Becher et al. 2004; Papoyan and Kochian, 2004, Küpper et al. 2007b). Strong sulfur ligands like phytochelatins were shown not to be relevant for cadmium detoxification in the Cd hyperaccumulator *T. caerulescens*. Phytochelatin levels are lower in this plant than in the related non-accumulator *T. arvense* (Ebbs et al. 2002), inhibition of phytochelatin synthase in hyperaccumulators does not affect their Cd resistance (Schat et al. 2002), and direct measurements of the Cd ligands by EXAFS showed that most of the Cd in this species is not bound by strong ligands, but by weak oxygen ligands (Küpper et al. 2004). Thus, the main detoxification strategy in hyperaccumulators is clearly not binding to strong ligands, but sequestration of the hyperaccumulated heavy metals. However, the non-proteogenic amino acid nicotianamine seems to play an important role in metal homeostasis of plants. According to several studies it binds iron, zinc and copper, mainly for long-distance transport in the vascular bundle (Stephan and Scholz, 1993; Schmidke and Stephan, 1995; Stephan et al. 1996, Pich et al. 1994; Pich and Scholz, 1996; von Wiren et al. 1999; Liao et al. 2000), and nicotianamine (NA) synthase has been shown to be highly overexpressed in hyperaccumulators compared to non-accumulator plants (e.g. Becher et al. 2004; Weber et al. 2008; van de Mortel et al. 2006, 2008).

Under metal-induced stress, the heavy metal accumulation pattern changes. Under such conditions, heavy metal (Cd, Ni) accumulation was enhanced in a few cells of the mesophyll (Küpper et al. 2000a, 2001). The same cells contained elevated levels of magnesium, which was interpreted as a defence against substitution of Mg^{2+} in Chl (Küpper et al. 1996, 1998, 2002) by heavy metals. Recently it was found that this heterogeneity of Cd accumulation is a transient phenomenon in *T. caerulescens* under Cd-toxicity stress, correlating with a heterogeneity in photosynthesis and disappearing when the plants acclimate to the stress (Küpper et al. 2007a). It was postulated that this transient heterogeneity constitutes an emergency defence against Cd toxicity by sacrificing a few mesophyll cells as additional storage sites until the metal sequestration in the epidermis is sufficiently upregulated. Further, this acclimation response showed that at least part of the Cd resistance of *T. caerulescens* is inducible (Küpper et al. 2007a).

While Cd resistance of *T. caerulescens* is strongly enhanced compared to related non-accumulator plants, *T. caerulescens* is known to be sensitive to copper to a similar extent as non-accumulator plants, limiting its application for phytoremediating soils with mixed contamination (Walker and Bernal, 2004; Benzarti et al. 2008).

In contrast to other metals, the speciation (redox state and coordination) of copper in plant tissues is poorly understood, despite the detailed structural and functional knowledge about numerous copper-dependent enzymes and copper chaperones (e.g. review of Pilon et al. 2006). This is most likely due to the difficulty of measuring copper speciation. The concentrations of copper in plant tissue are too small for NMR and were too small for EXAFS for a long time. Methods relying on fractionation or homogenisation of fresh tissues (e.g. for chromatography) cause breaking of intracellular membranes. This brings weakly bound metal that was localised in the vacuole into contact with the various strong ligands of the cytoplasm, causing artefactual changes of speciation. To the authors' knowledge, so far only one EXAFS study on copper speciation in plant tissues used environmental copper concentrations: Polette et al. (2000) studied copper speciation in the copper resistant copper indicator plant *Larrea tridentata*. In their EXAFS measurements on samples frozen in liquid nitrogen. The authors found evidence for a copper(II)-phytochelatin complex involved in transport and an unknown copper complex involved in copper storage. A second study using the same sample preparation method (that due to the formation of a gas layer around the sample slows down freezing, resulting in ice formation with the risk of membrane damage) investigated the Cu-tolerant Cu-excluder plant *Elsholtzia splendens* (Shi et al. 2008). Unfortunately, these authors applied very high, physiologically not relevant copper levels (300 µM) in their nutrient solution in order to force enough copper into the tissues to get an acceptable signal/noise level in their EXAFS data within the limited synchrotron beam time. In two further copper EXAFS studies on plant tissues (Sahi et al. 2007 on *Sesbania drummondii*; Gardea-Torresdey 2001 on *Larrea tridentata*), strong artefacts were most likely introduced by drying of tissues. Drying disrupts membranes as discussed above, and can remove aquo ligands from metal ions (e.g. Schünemann et al. 1999). Additionally, extremely high copper concentrations in the nutrient solution were applied to the plants (1000 and 10,000 µM by Gardea-Torresdey et al. 2001; 400-5,000 µM by Sahi et al. 2007). Thus, the biological relevance of these studies has to be questioned.

In the current study, we addressed two related questions: how do environmentally relevant (10 µM) toxic levels of copper affect *T. caerulescens*, and how does the plant try to defend itself by binding the copper ions, both in comparison to the copper accumulator *C. helmsii* described in our companion paper. This was done by investigating mechanisms of copper-induced inhibition of photosynthesis, and analysing the complexation of copper in leaves. The latter was done in order to find out whether *T. caerulescens* utilizes different mechanisms for detoxification of the non-accumulated copper in comparison to the hyperaccumulated cadmium. The speciation of copper was analysed by extended X-ray absorption fine structure (EXAFS), which is an element specific method and therefore particularly suited for analysing the *in vivo* ligand environment of metals in plants. Furthermore, in contrast to most other methods it is applicable to intact frozen-hydrated plant tissues, which has first been done by Salt et al. (1995). Moreover, these days EXAFS beamlines can reach much lower metal concentrations than heteronuclear NMR (Ascone et al. 2003). So far, metal NMR was applied only in the case of cadmium in *T. caerulescens* (Ueno et al. 2005), but although the concentration of this hyperaccumulated metal in the tissue was about hundred times higher than that of copper now, the NMR signal obtained was so weak that only the most abundant ligand type (organic acids, confirming previous EXAFS work, Küpper et al. 2004) could be detected, everything else drowned in noise. Thus, EXAFS is the method of choice for identifying ligands of both Cu(I) and Cu(II) in intact tissues in the absence of artefacts introduced by sample preparation (see above). EXAFS model complexes were prepared as references, and among these, Cu(II)-nicotianamine was characterised in more detail by UV/VIS and EPR spectroscopy in view of its importance for Cu(II)-binding in *T. caerulescens*. Inhibition of photosynthesis was investigated, as in our *Crassula helmsii* study, by the two-dimensional (imaging) microscopic *in vivo* measurement of the transients of chlorophyll variable fluorescence with a recently described instrument, the Fluorescence Kinetic Microscope (Küpper et al. 2007a).

This was complemented with data on copper accumulation, changes in pigment composition and plant growth.

RESULTS

Plant Growth and Metal Uptake, Visual Symptoms of Copper Stress

There was little difference in growth inhibition by copper between the Cd/Zn-hyperaccumulator *Thlaspi caerulescens* and the related Cd/Zn non-accumulator (possible Ni-accumulator) *T. fendleri*. At 10 µM Cu^{2+} for eight weeks, fresh weight of *T. caerulescens* was reduced to 1% of the control, *T. fendleri* to 1.5% (Table 1). Remarkably, in this case the hyperaccumulator *T. caerulescens* was even slightly more sensitive than the non-accumulator *T. fendleri*. A few individuals of *T. caerulescens*, however, turned out to be much more copper resistant, which became obvious mainly after a longer time of growth, indicating acclimation. After 3-4 months of growth, the copper-resistant individuals had, on average, about 3 times higher fresh weight compared to the regular sensitive individuals (Table 1). This was not related, however, to some physiological metal exclusion mechanism, less supply of nutrient solution on the pot (as resistant and non-resistant plants were growing in the same pots), or any other effect reducing copper load in the shoots. Surprisingly, the copper content in leaves of the resistant individuals was even 25% higher than in leaves of sensitive individuals (Table 1). Nevertheless, the copper contents in all *T. caerulescens* leaves were typical for non-accumulator plants (less than 50 ppm). The resistant individuals looked much healthier in terms of the number and size of living leaves, but also they had slightly chlorotic leaves (Figure 1 shows one of the most Cu-resistant individuals we observed).

Fig. 1. Visual differences after growth in hydroponic solution for 4.5 months. Top: two plants stressed in the same pot with, 10 µM Cu^{2+}, showing the difference in growth between a copper sensitive (left) and a copper resistant (right) individual of *Thlaspi caerulescens* (Ganges population). Bottom: control plant grown at 0.1 µM Cu^{2+}.

Some of the Cu-stressed plants stayed green, others became partially chlorotic. The first case was correlated with the occurrence of some [Cu(II)]-Chl a in extracts of the plants, although the [Cu(II)]-Chl content never exceeded 2% of total Chl. This and the bleached plants showed that the type of damage in these plants was closer to the "sun reaction" (characterised by heavy metal insertion only in the PS II reaction centre) than to the "shade reaction" (characterised by the formation of heavy metal substituted chlorophylls in the LHCII antenna of PSII, Küpper et al. 2002). The bleaching in this case cannot be due to bleaching of [Cu(II)]-Chl because this pigment is highly stable (Küpper et al. 1996).

Cu-induced inhibition of Photosynthesis

The biophysical meaning of the fluorescence kinetic parameters used in the following text is described in more detail in the list of abbreviations, an introduction into the subject is given e.g. by the review of Maxwell and Johnson (2000). Copper stress in *T. caerulescens* decreased the maximal dark-adapted quantum yield of PSII photochemistry, $F_v/F_m = (F_m-F_0)/F_m$, as well as the light-acclimated efficiency of electron transport through PSII, $\Phi_{PSII} = (F_m'-F_t')/F_m'$ (Fig. 2). This was caused by a dramatic increase in the basic fluorescence yield F_0 that was stronger than the increase of the maximal fluorescence yield F_m. While Cu-toxicity increased F_0 slightly more (410% instead of 360%) in the Cu-sensitive than in the Cu-resistant individuals, F_m increased more (about 20% instead of 10%) in the latter. Based on the above mentioned definitions, this led to a milder decrease of F_v/F_m and Φ_{PSII} in the Cu-resistant compared to the Cu-sensitive specimens. The values of non-photochemical quenching of excitation energy (NPQ = $(F_m - F_m')/F_m$)) became negative under copper stress, i.e. more excited states of Chl relaxed via fluorescence in the light-adapted compared to the dark-adapted state of the photosynthetic system. In the Cu-sensitive individuals these negative NPQ values were of similar amplitude as the positive NPQ values in the control. In the Cu-resistant individuals, NPQ was almost zero. The light saturation of the photosynthetic system, measured as $(F_p-F_0)/(F_m-F_0)$ as described before (Küpper et al. 2007a), was very strongly affected by copper. Light saturation of the copper-stressed plants was about 2 for the sensitive and 1.4 for the resistant copper-stressed plants, which means that the maximal fluorescence quantum yield after onset of actinic light (F_p) was higher than the normally highest maximum fluorescence quantum yield, which is measured in the dark-adapted state (F_m). In contrast, light saturation was only 0.25 for the controls (Fig. 2).

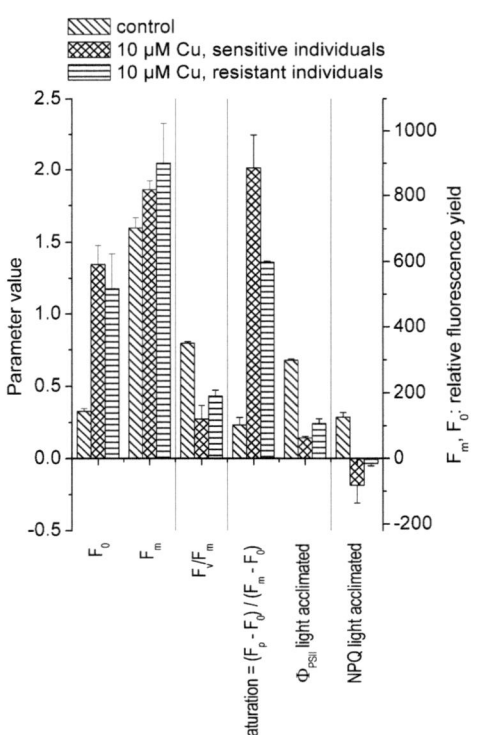

Fig. 2. Fluorescence kinetic microscopy of copper-stressed *Thlaspi caerulescens*. The data are from a typical experiment from the second series of experiments (see methods), the same experiment as used for Fig 3 of our publication about Cd acclimation (Küpper et al. 2007a) so that the controls were the same and the effects of Cu (reported now) and Cd (reported before) are directly comparable. Actinic irradiance during measurement: about 40 $\mu mol.m^{-2}.s^{-1}$. Averages and standard errors of all measurements between 3 and 5 weeks of treatment with copper. The meaning of the fluorescence kinetic parameters is explained in the list of abbreviations.

Characterisation of copper ligands by EXAFS, EPR and UV/VIS spectroscopy

Like in our previous EXAFS studies including the companion paper on *Crassula helmsii*, we characterised several copper complexes in solution. They served as a basis for evaluating the EXAFS spectra of plant samples, where many ligands could contribute to copper ligation. These model complexes were selected because they are likely copper ligands themselves (Cu(II)-histidine, Cu(II)-nicotianamine), or because they could be used as simplified models of complex ligands *in vivo* (glutathione as a first-shell ligand model for the various sulfur ligands in plants).

Cu(II)-histidine

A fit of its EXAFS spectrum with four histidines bound by the imidazole nitrogen had the best fit index (FI), but the Debye-Waller factors (DW) of the higher shells were rather high, indicating larger than expected disorder in those ligand shells. Thus, fits with lower coordination of histidine and addition of a low Z ligand (oxygen) were tried. The best combination of FI and DW parameters was found for two histidines (bound by the imidazole nitrogen) and two oxygens, while 6-coordination models resulted in unrealistically higher DW parameters (Fig 3).

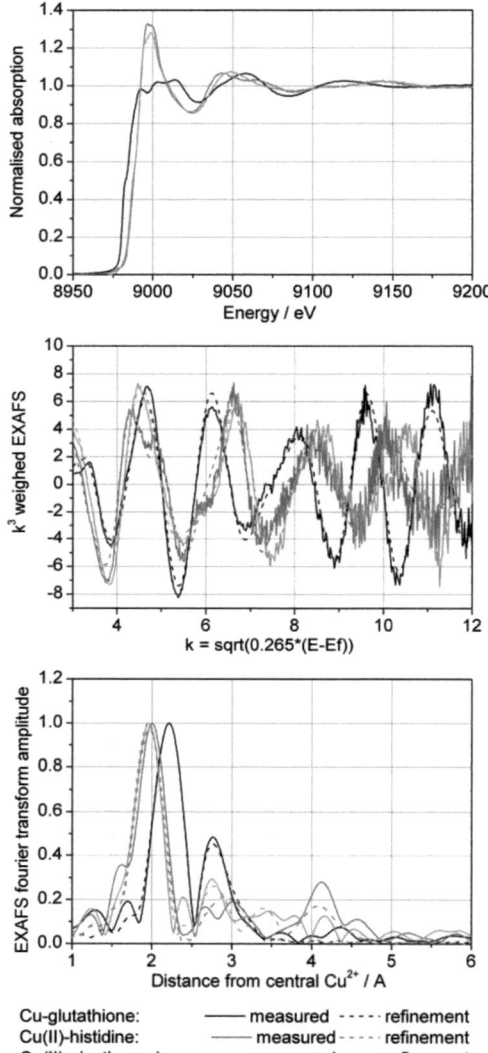

Fig. 3. Comparison of Cu X-ray absorption spectra of different model compounds. ***Top:*** Raw data after normalisation. Note that X-ray absorbance was measured as an excitation spectrum of X-ray fluorescence. ***Middle:*** Normalised EXAFS. ***Bottom:*** Fourier transform of the EXAFS.

Cu(II)-proline

In the case of Cu(II)-proline, the best fit was for 4-coordination of low Z ligands (oxygens were used). The higher shell contributions could not be fitted with imidazole rings (emulating the proline ring structure), which may indicate that proline was bound not by the ring but by the carboxylate end.

Cu(I/II)-glutathione

This was a mixture of the Cu(I) and the Cu(II)-complex as generated by preparation method. Its EXAFS spectrum (Fig 3) was modelled with copper site coordinated to 4 sulfur atoms in the first shell. A second shell of carbon led to higher FI than a Cu-Cu interaction at about 2.7 Å. For a mixed first shell with sulfur and low Z ligands (oxygen was tried) the best fit was found for 3 sulfur and 1 oxygen and a Cu-Cu interaction in the higher shell. However, this model was discarded because of the very long (unrealistic) distance obtained for the bound oxygen. The DW parameter found for oxygen was small for a 4-coordinated copper complex, suggesting that the coordination number might be higher. However, increasing the coordination to 5 by adding one more oxygen led to a negative DW parameter for that oxygen. In summary, the most realistic model consists of a CuS_4 center and a Cu-Cu interaction due to formation of a cluster with two coppers bridged by sulfurs, plus coordination of each copper by another two sulfurs from glutathione (i.e. Cu_2GS_6).

Cu(II) nicotianamine

EXAFS. The results were similar for both pH values analyzed (pH 4 and pH 7). For modelling the structure of nicotianamine in the refinements, we used the mugeinic acid structure published by Nomoto et al. (1981) because both molecules are almost identical. For both nicotianamine samples the fits including four ligands in the first shell and two additional long ligands gave a rather high DW parameter for the long ligands. The improvement in the FI was tiny, which with an increased number of parameters reduces the significance of the additional ligands. Therefore, the EXAFS result alone was not sufficient to decide whether the Cu(II) bound to nicotianamine was 4- or 6-coordinated. However it is well in agreement with a 6-coordinated model where two highly mobile water molecules are bound at longer distances (Table II, Fig. 3).

EPR spectra. In order to obtain a more definite answer about the coordination of copper in the Cu(II)-nicotianamine complex, EPR spectroscopy was applied. At X-band, 20 K, Cu(II)-nicotianamine exhibited a rather simple close to isotropic signal over the range pH 3-7, with a baseline crossing at $g \approx 2.12$ (Fig.4b). The signal kept its shape independent of temperature (5-70 K) and concentration of copper (0.43 - 0.10 mM). Cu(II)-nicotianamine was fully EPR active under our experimental conditions as documented by double integration of the signal, and there were no transitions detectable at half-field indicative for Cu-Cu interaction. Overall, the EPR properties of Cu(II)-nicotianamine are in line with a hexadentate complex, similar in structure to the related Cu(II)-mugineic acid complex whose three-dimensional structure has been obtained by X-ray crystallography (Nomoto et al. 1981). Nicotianamine provides six alternating carboxylate and amine functions whose relative positions favour the formation of six-coordinate metal complexes as discussed most recently (Callahan et al. 2006).

UV/VIS-Spectroscopy.

The absorption spectra of nicotianamine showed an absorption maximum at 272 nm (Fig. 4a). Addition of Cu^{2+} at pH values above 2 caused a pronounced shift of the maximum to 247 nm and a very strong increase in intensity of this band. Furthermore, a weak additional absorption maximum appeared at 609 nm ($\varepsilon_{609} \approx 100\ M^{-1}cm^{-1}$). At pH = 2, the spectrum of the nicotianamine+Cu^{2+} mixture already had a shape intermediate between the spectrum of nicotianamine alone and that of the Cu(II)-nicotianamine complex at higher pH values: it had a low peak at 247 nm and a residual shoulder at 272 nm. Also the 609 nm absorption peak started to appear at pH 2 (Fig. 4a).

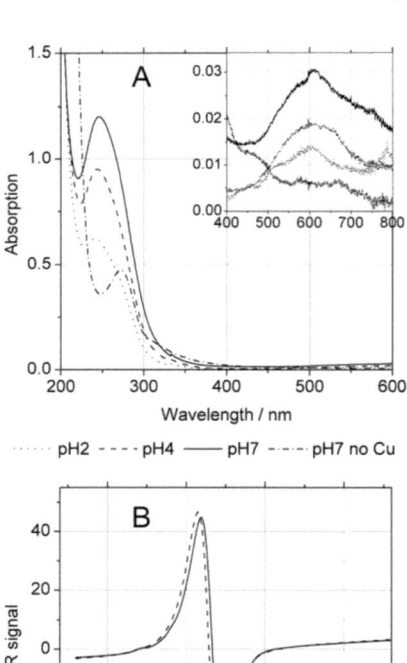

Fig. 4 Optical and Electron Paramagnetic Resonance spectra of Cu(II)-nicotianamine.

A) UV/VIS-Absorption spectra of 0.4 mM nicotianamine or 0.4 mM nicotianamine with 0.32 mM Cu^{2+} ($CuCl_2$) in the short (left) and long (right) wavelength range.

B) EPR-Spectra of the nicotianamine copper complex at pH 2-7.

EXAFS spectroscopy of plant samples

In contrast to typical EXAFS spectra on metals in biological systems, here more than a single contribution (or metal to neighbour atom distance) could be identified, which was visualized by corresponding peaks in the Fourier transforms. Generally, the EXAFS spectra of all plant samples had a much higher noise, and therefore worse statistics in particular in the high k-range (i.e. high above the copper absorption edge), compared to the model compounds (Fig. 5). This noise level was inevitable because *T. caerulescens* is not a copper hyperaccumulator (see AAS data), and environmentally relevant copper concentrations have been used for plant growth. Therefore, we had to measure each sample for about three days until the statistics was sufficient for detailed analysis. Thus the current set of data, consisting of 8 plant samples, amounted to almost a month of EXAFS beamtime. For getting a further reduction of the noise to half, four times more integration time (12 days per sample) would have been needed, which would have been impossible at any synchrotron beamline. Like the refinements, also the component analysis was limited by this noise level, but still it yielded reasonably reproducible data as indicated by the error bars after averaging the data from independent plant samples (Fig. 6). As malate, citrate and the aquo complex could not be distinguished from each other, their values in the fits are presented only as the sum of these three components.

For the first shell contributions in the refinements (data shown in Table 3), oxygen/nitrogen and sulfur ligands were tested. For contributions from higher shells, nicotianamine was tested because it was previously predicted to play a role in copper metabolism (Pich & Scholz 1996; Liao et al. 2000; Irtelli et al. 2009). Nicotianamine was fitted using the parameters (distances, ligand types and numbers of the individual ligand shells) of the model compound. As Cu(II)-nicotianamine is a metal:ligand 1:1 complex, this structure should not be different in the presence of other potential ligands. The overlap of the long first shell oxygen ligands with possible sulfur or phosphorus contributions was taken into account in the interpretation of the data. Interestingly, nicotianamine alone could not explain the high amplitude of the peak at 2.7-3 Å; tests of possible ligand atoms resulted in the conclusion that phosphorus or sulfur is most likely the second contributor to this peak. The contributions at above 3.3 Å could not be explained by nicotianamine, histidine or any other of the tested model compounds. Various atom types were tested for each major contribution, resulting in the conclusion that these contributions show copper-copper interactions from rigid copper clusters. It should be noted that this contribution is so dominant at higher k-values in the EXAFS that even at the present noise level the identification of these metal clusters is significant.

Besides these general observations, the following sample-specific characteristics of Cu-ligation were found.

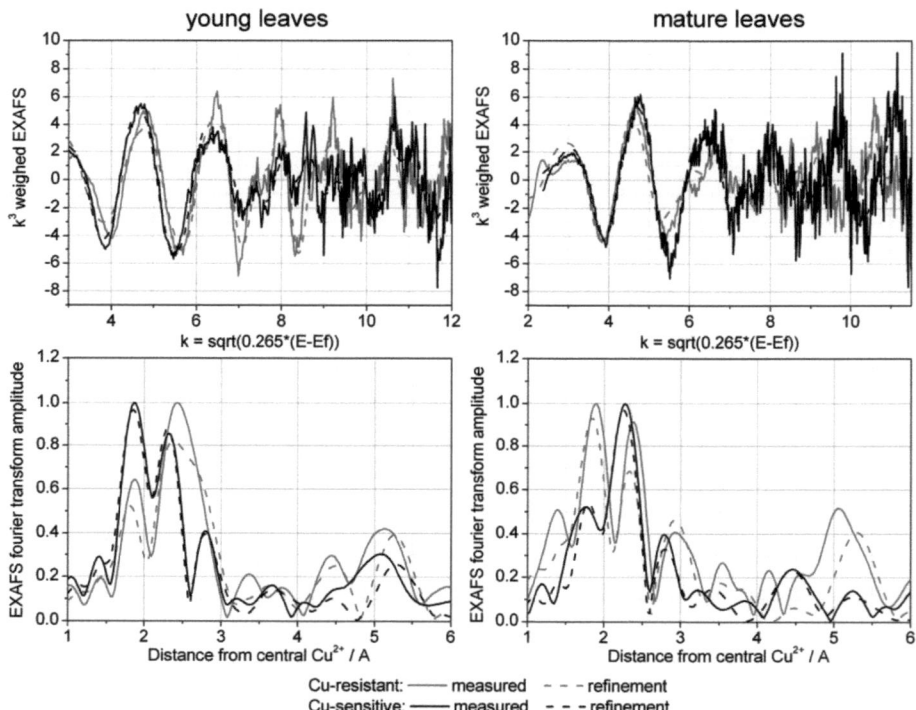

Fig. 5. Typical examples of data obtained by *in situ* X-ray absorbance measurements (Cu K-edge) of frozen-hydrated *Thlaspi caerulescens* tissues. This example represents a measurement of Cu-EXAFS in a mature leaf of a Cu-resistant specimen. ***Top:*** Normalised EXAFS and fit with theoretical model. ***Bottom:*** Fourier transform of the EXAFS and fits with theoretical model (done on normalised EXAFS) as well as fit with datasets of model compounds.

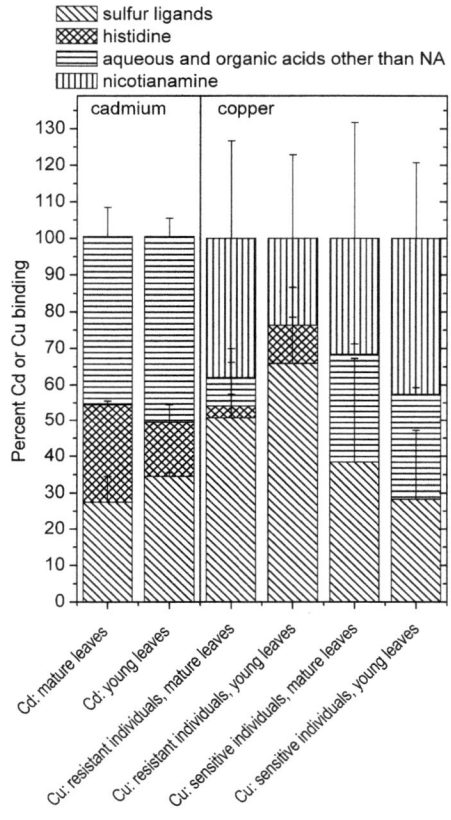

Fig. 6. Metal, age and specimen-dependent differences in ligand environment. **Right:** Averaged samples of leaves from plants grown on 10 µM Zn^{2+} and 10 µM Cu^{2+}: Cu K-edge (data from current study). Plant EXAFS spectra (k^3-weighed) were fitted with a linear combination of the EXAFS spectra of the following model compounds: aqueous ($CuSO_4$), Cu(II)-malate, Cu(II)-citrate, Cu(II)-proline, Cu(II)-histidine, Cu(II)-nicotianamine pH4, Cu(II)-nicotianamine pH7, Cu(I/II)-glutathione. Proline was never detected as a ligand, as was the aquo complex. The data of Cu(II)-malate, Cu(II)-citrate were averaged like the data of Cu(II)-nicotianamine at pH4 and pH7. **Left:** Averaged samples of mature leaves from plants grown on 10 µM Zn^{2+} and 100 µM Cd^{2+}: Cd K-edge (data from our previous study: Küpper et al. 2004). Plant XAS spectra were fitted with a linear combination of the EXAFS spectra of the following model compounds: aqueous ($CdSO_4$), Cd(II)-malate, Cd(II)-citrate, Cd(II)-histidine, Cd(II)-glutathione. The data of Cd(II)-malate, Cd(II)-citrate and the aquo complex were averaged. Similar results as for Cd were obtained for Zn, see Küpper et al. (2004).

Cu-resistant plants, young leaves (3 samples):

Copper was found to be four-coordinated in the first ligand shell, with a mixture of oxygen and sulfur ligands. On average, 1.7 (± 0.2) sulfur ligands were found. Further, 2.2 (± 0.2) oxygen ligands were found in the first ligand shell, of which 1.6 (± 0.7) were ascribed to oxygen ligands from nicotianamine carboxylates by the characteristic contribution at 2.8 Å (see model compound) that is not detectable in small organic acids like malate or citrate. In one of the samples, a fit with histidine instead of nicotianamine yielded a better DW parameter, probably because in this sample also the characteristic scattering of nicotianamine at 3 Å was less pronounced (not shown). Besides nicotianamine, the 3 Å peak included contributions from phosphorus or sulfur – fitted as 2.6 (±0.3) phosphorus atoms in the refinements shown in table 3. Further, all these samples showed a particularly intense peak in the Fourier transform at rather high distance (about 5.2 Å), which was best fitted by a copper-copper interaction with 4.5 (±0.8) Cu at this distance from the central atom (Fig. 5). Also the peaks between this and the 2.7 - 3 Å peak could only be explained by a heavy backscatterer like copper. In summary, it can be clearly said that oxygen and sulfur ligands contributed to the copper binding in the first shell, and there is evidence for contributions from nicotianamine and copper-copper interactions. Possibly histidine was involved as a ligand as well. This result was confirmed by the component analysis, where around 65% of the copper ligands were found to be sulfurs, with the remainder being mainly histidine (Fig. 6).

Cu-resistant plants, mature leaves (2 samples):
Similar results as for the young leaves were also obtained for the mature leaves. Again, 1.6 (±0.4) sulfur ligands and 2.2 (±0.4) oxygen ligands were detected in the first shell. The oxygen ligands were almost all (2.1 ±0.4) identified as nicotianamine by its characteristic contribution at about 2.8 Å (see above). In the best data set, a reasonable signal/noise ratio was available till $k=11.5$ $Å^{-1}$. When using histidine as a ligand, the best fit was found for 2 histidines and 2 sulfurs bound to Cu. But also in this sample, the fit including nicotianamine instead of histidine gave a better picture, because this spectrum had a rather sharp peak at about 3 Å in its Fourier transform, which characteristic for nicotianamine (see above). In the outer shells, again strong copper-copper contributions were found, with 3.1 (±0.6) Cu at 5.2 Å from the central Cu atom. The component analysis supported the contribution of nicotianamine rather than histidine to copper binding. Further, it clearly showed a strong sulfur contribution, but slightly less sulfur than in the young leaves (50% instead of 65%, Fig. 6).

In summary, it seems that also in mature (like in young) leaves of resistant individuals of *T. caerulescens* copper is mainly bound by a combination of sulfur (thiolate) ligands and nicotianamine, with a noticeable shift from sulfur ligands to nicotianamine coordination during leaf maturation. Like in the mature leaves, the strong contributions at very large distances from the central copper could be explained only by assuming a rigid cluster of copper so that copper-copper-contributions become visible in the EXAFS spectrum.

Cu-sensitive plants, young and mature leaves (3 samples)
For the first shell the best refinement was obtained for 1 (±0.3) sulfur and 2.9 (±0.4) oxygen; of the oxygens 2.2 (± 0.9) were refined as nicotianamine. Like the sulfur contribution in the first shell, also the sulfur/phosphorus contribution at about 2.9 Å was less pronounced than in the Cu-resistant

individuals – 0.7 (±0.2) P were refined for the sensitive individuals compared to 1.9 (±0.3) for the resistant individuals. As in the samples of the resistant plants, again there was a large peak at 5.2 Å in the Fourier transform that was ascribed to a copper-copper contribution. However, it was smaller than in the resistant plants; on average, 4 (±0.8) copper atoms were refined in the 5.2 Å peak in the Cu-resistant individuals, while 3.1 (±0.3) copper atoms were refined for this peak in the Cu-sensitive individuals. The component analysis again confirmed the refinements by yielding a mixture of sulfur and O/N ligands, but additionally showed better the contribution of nicotianamine (Fig. 6). While the proportion of copper binding to nicotianamine was similar as in the resistant individuals, the sulfur contribution was clearly lower, and a rather strong contribution from hydrated ions and/or from organic acids such as malate or citrate was found instead.

DISCUSSION

Here, we analysed the response of the Cd/Zn model hyperaccumulator *Thlaspi caerulescens* to the non-accumulated metal copper, to which this species is as sensitive as related non-accumulator species. This sensitivity is important because it limits the application of *T. caerulescens* for the phytoremediation of soils with mixed contamination, as recently tested (Walker and Bernal, 2004). Our study has revealed that this may be due to the metal specificity of the metal detoxification system in *T. caerulescens*, since copper was bound by completely different predominant ligands than cadmium or zinc. This is particularly interesting because in our companion study on copper metabolism in *Crassula helmsii* we demonstrated that in this copper accumulating species this metal was bound to weak oxygen ligands similar to the binding patterns for cadmium and zinc in *T. caerulescens*. Further, in the current study, in all experiments a few individuals turned out to be somewhat more copper resistant than others, and notably, those resistant individuals again had different copper ligands.

In addition to new insights into the ligand environment of copper in *T. caerulescens*, in our assessment of the vitality of the plants by measuring *in vivo* Chl fluorescence kinetics we found interesting aspects of the mode of inhibition by this metal. In particular, we observed an enhanced coupling of LHCII to PSII, most likely caused by copper-induced inhibition of the water splitting complex.

Physiological effects of copper on sensitive and resistant individuals

Like in our companion paper on copper effects in *Crassula helmsii*, in this study we investigated copper stress under high-irradiance conditions or with long light phases, leading to the "sun reaction" type of damage to photosynthesis (Küpper et al. 1996, 1998, 2002) with an inhibition of the photosystem II (PSII) reaction centres, like due to insertion of Cu^{2+} into the pheophytin a of the PSII reaction centre (Küpper et al. 2002). The resulting decrease of F_v/F_m was observed also by Lanaras et al. (1993) with Cu-stressed *Triticum aestivum*. This decline resulted from an increase of the basic fluorescence yield F_0 and a decrease of the variable fluorescence yield F_v in line with both, our former studies and those of Lanaras et al. (1993). The same applies to the decline of the quantum yield of PSII photochemistry in light acclimated state as measured by the Φ_{PSII} parameter. The increase of F_0 despite the decreasing pigment content of the leaves looks surprising at first glance, but was observed earlier (Ouzounidou et al. 1997, Küpper et al. 2007a) and is most likely caused by deeper penetration of exciting light and fluorescence emission from deeper leaf layers. The stronger decrease of Φ_{PSII} compared to F_v/F_m indicates the presence of a second inhibition target in addition to the inactivation of PSII reaction centres, such as inhibition of electron transfer after PSII or inhibition of the water splitting complex as postulated earlier. The increasing light saturation under copper stress observed now indicates that in a system of high antenna connectivity the number of functional PSII reaction centres decreases faster than the number of functional antenna complexes; similar effects were observed for cadmium stress in *T. caerulescens* (Küpper et al. 2007a). While under Cd stress, however, the non-photochemical quenching of excitation energy increased, copper stress led to negative values of $NPQ = (F_m-F_m')/F_m$. Such negative NPQ values, meaning that maximal fluorescence quantum yield is larger in light acclimated than in dark acclimated state, are possible only if an increase of the PSII-associated antenna occurs in response to light. Hence, it may be concluded that copper causes enhanced state 1 transitions, i.e. migration of LHCII from PSI to PSII. This makes sense if the second point (besides the PSII reaction centre) in the inhibition of photochemistry is an inhibition of the water splitting complex. Such an inhibition would result in a strongly oxidised plastoquinone pool since PSII could "pump" as many electrons into it as PSI takes out of it, and an oxidised plastoquinone pool is well-known to induce state I transitions.

All copper toxicity induced effects discussed above were milder in the few copper-resistant compared to the normal copper-sensitive individuals of *T. caerulescens*. Interestingly, the less severe reduction of F_v/F_m and Φ_{PSII} was only partially caused by a lesser increase of F_0, but also by a stronger increase of F_m in the resistant compared to the sensitive individuals. This indicates that the resistance did not only involve a better copper detoxification (see discussion of Cu-ligands below), but also an enhanced repair of chlorophyll-protein complexes.

Cu-ligands in Thlaspi caerulescens

Metallothioneins and Phytochelatins

Clearly, a strong contribution of sulfur ligands to binding of copper was found in most of the samples. In view of a generally very high expression of metallothioneins in *T. caerulescens* (Papoyan and Kochian 2004), it is most likely that this sulfur binding at least partially originates from Cu(I)-S clusters in metallothioneins. Already previous studies indicated that in plants metallothioneins are much more relevant for copper homeostasis than for zinc homeostasis (Cobbett and Goldsbrough 2002), and physiological studies suggested a role for them in *T. caerulescens* (Roosens et al. 2004). Further, copper(I) metallothioneins are known to form very rigid copper-sulfur clusters, that stabilise copper(I) in aqueous solution (Byrd et al. 1988). Strong multiple scattering due to this rigid structure is to be expected (Calderone et al. 2005), and might contribute to the EXAFS peak at 3.7 Å (George et al. 1988). A contribution from metallothioneins would further explain why the XANES (the x-ray absorption close to the k-edge) of our spectra shows a clear contribution of monovalent copper despite the high oxygen content in plant tissues. All these indications of a metallothionein contribution were much more pronounced in the copper-resistant compared to the copper-sensitive individuals of *T. caerulescens*. In contrast, copper phytochelatins have a completely different EXAFS spectrum, with only one strong Cu-Cu interaction peak at 2.56 Å and a weak contribution at 4.04 Å (Polette et al. 2000), but no contribution at the distances where we observed peaks in the Fourier transforms of our samples (2.8 – 3 Å, 3.7 Å, 4.4Å and 5.2 Å). Therefore, we can conclude that phytochelatins do not bind a major proportion of copper in *T. caerulescens* leaves..

Biomineralisation

Modelling our EXAFS spectra with contributions from metallothioneins could explain the sulfur contribution to the first ligand shell and possibly a copper-copper interaction at 3.7 Å. But is does not explain the other contributions, e.g. the strong peaks at about 2.8 – 3 Å, 4.4Å and 5.2 Å. Like the metallothioneins, the latter two contributions occurred in particular in the more Cu-resistant individuals of *T. caerulescens* and could be identified for all samples. Their high intensity (in particular of the 5 Å peak) suggests it is a metal-metal interaction, because lighter elements cannot result in such a large peak at such an extreme distance from the central atom. Even multiple scattering within the first shell that recently has been identified in a very limited number of metalloproteins (Hollenstein et al. 2009, Ha et al. 2007) would occur at shorter distances and thus can be ruled out. The identity of the metal at 5 Å can be obtained from the EXAFS data only with an uncertainty of one or two atomic numbers, but due to the metal content in the plant samples it could only be a Cu-Cu interaction. Comparing our data with literature showed a striking similarity of these and other scattering peaks found in our samples with copper moolooite, a hydrated crystalline form of copper oxalate. EXAFS of copper oxalate was described already in 1979 by Michalowicz et al. and it was first described as a naturally occurring mineral, moolooite, by Clarke and Williams (1986). Soon after, it was detected to occur in copper-tolerant lichens (Chisholm et al. 1987; confirmed by Purvis et al. 2008), and recently it was shown (by Cu-EXAFS) to be formed by the highly heavy metal tolerant fungus *Beauvaria caledonica* (Fomina et al. 2005). The Cu-EXAFS spectra published by the latter authors clearly showed all characteristic peaks of Cu-moolooite in the fungal "biominerals". In our own data, qualitatively all contributions to be expected from Cu-moolooite were present as well. The 5 Å peak, however, was often much higher (relative to the other contributions) in our samples than what could be expected from the spectrum of Cu-moolooite. Therefore, it cannot be excluded that another substance contributed to the 5 Å Cu-Cu interactions in *T. caerulescens*. Only few compounds are known to have a high contribution at such an extreme distance from the central atom. Candidates are a copper oxide, CuO, also known as the mineral tenorite (EXAFS spectrum published e.g. by Zhou et al. 1999), and hydrocalcite (EXAFS spectrum published e.g. by Porta et al. 1996). Features similar to the ones observed in our *T. caerulescens* samples were reported for samples collected from *Larrea tridentata* in its natural environment (Polette et al. 2000), but at that time no reference compounds with similar Cu-EXAFS spectrum were known. Therefore, these authors called this copper speciation "unknown", with an assumption based on non-EXAFS data about bond distances that the 3.7 A peak in the EXAFS spectrum may be attributed to Cu-Cu-interactions in chalcopyrite. Further, as their samples were collected in the natural environment and contained copper-rich dust particles inside the stomatal cavities, in their case this copper-mineral was not due to bio-mineralisation. In our case, such deposits of dust from copper minerals can be excluded as our plants were raised in a growth chamber on hydroponic solution. Revealing the exact identity of the copper clusters that give rise to the extremely intense features in the EXAFS spectra of *T. caerulescens* will be an exciting but challenging task for future research.

Nicotianamine

Most of the remainder of the higher shell contributions, in particular the peak at 2.7 – 3 A, seems to be due to copper binding to the non-proteogenic amino acid nicotianamine, which has been shown to have a very high affinity for copper (Beneš et al. 1983) and already earlier works suggested, on a physiological basis, that nicotianamine is involved in copper homeostasis (Pich & Scholz 1996; Liao et al. 2000; Irtelli et al. 2009). The explanation of the features in our EXAFS spectra with carboxylate-bound nicotianamine seems more likely than explanation by histidine, as seen by the statistical analysis of the fits. Because of the finding that nicotianamine plays a role in copper binding in *T. caerulescens*, we further characterised this ligand in isolated form. The combination of EXAFS and EPR data of Cu(II)-nicotianamine showed that Cu(II) was coordinated in a hexadentate way, with four short-distance and two long-distance ligands. This is in agreement with the structure of the mugeinic acid model from Nomoto et al. (1981). Titration indicated that only above pH 4 copper binds to the carboxyl groups of NA. The UV/VIS absorption and EPR spectra at different pH values, however, strongly change already above pH 3. Also the Cu-NA EXAFS spectra at pH 4 and pH 7 yielded similar results, showing the scattering pattern typical for NA. Obviously the latter methods are better in detecting the beginning of the copper binding to nicotianamine, where the interaction is not yet strong enough to influence the pH value.

Summary of copper complexation strategies in *T. caerulescens*

The finding of strong contributions from sulfur ligands and from the tightly binding amino acid nicotianamine, as well as the discovery of copper clusters that strongly suggest the occurrence of copper biomineralisation in *T. caerulescens*, was surprising in view of our earlier study (Küpper et al. 2004) on ligands of the hyperaccumulated Cd and Zn in the same species. That study had shown that in these plants zinc and even the normally sulfur ligand-bound cadmium is bound by weak oxygen/nitrogen ligands. The new finding of predominantly strong ligands around copper indicates that hyperaccumulators have completely different strategies of detoxification for metals that are hyperaccumulated (in this case Cd and Zn) compared to non-hyperaccumulated metals (here copper). For the hyperaccumulated metals, detoxification is mainly based on active sequestration into the vacuoles of the epidermis, where they are stored only loosely associated with organic acids that are anyhow abundant in this organelle. Strong ligands like the phytochelatins and metallothioneins that detoxify heavy metals in non-accumulator plants do not play a major role in the detoxification of hyperaccumulated metals in hyperaccumulator plants. This view was now reinforced also by our companion study on copper metabolism in *Crassula helmsii*, where copper was bound exclusively by weak oxygen ligands (most likely organic acids such as malate). In contrast, it seems that hyperaccumulators deal with non-accumulated metals in the same way as non-accumulator plants, i.e. by binding them with strong ligands like metallothioneins and nicotianamine. In addition, strong indications were found for a detoxification of copper by biomineralisation. These forms of copper detoxification, in particular metallothioneins and biomineralisation, are clearly enhanced in the copper-resistant compared to the copper-sensitive *T. caerulescens* individuals. Therefore, phytoremediation of soils with mixed contamination including copper could most likely be improved (compared to the normal wild-type population of the Ganges ecotype) by selective breeding of the Cu-tolerant *T. caerulescens* individuals.

MATERIAL AND METHODS

Plant material, culture media and culture conditions

Seeds of *Thlaspi caerulescens* J.&C. PRESL (Ganges population), *Thlaspi fendleri* (NELS.) HITCHC and *Thlaspi ochroleucum* BOISS ET HELDER were germinated on a mixture of perlite and vermiculite moistened with deionised water. Three weeks after germination, seedlings were transferred to a nutrient solution containing 1000 µM $Ca(NO_3)_2$, 500 µM $MgSO_4$, 50 µM K_2HPO_4, 100 µM KCl, 10 µM H_3BO_3, 0.1 µM $MnSO_4$, 0.2 µM Na_2MoO_4, 0.1 µM $CuSO_4$, 0.5 µM $NiSO_4$, 20 µM Fe(III)-EDDHA (Fe(III)-ethylenediamine-di(o-hydroxyphenylacetic acid), and 10 µM $ZnSO_4$ (i.e. like Shen *et al.* 1997, but lower Cu and Mn). The pH of the solution was maintained at around 6.0 with 2.0 mM MES (2-morpholinoethanesulphonic acid, pH adjustment with KOH). As in many previous studies (e.g. Küpper et al. 1999, 2004, 2007a, 2007b, Lombi et al. 2000) the nutrient solution contained 10 µM Zn because of the high Zn requirement of Zn hyperaccumulators (e.g. Shen et al. 1997). The nutrient solution was aerated continuously. After 12±2 days of growth, 10 µM Cu^{2+} (as $CuSO_4$) was added to half of the pots. We used this concentration because initial tests had shown that it is sublethally toxic to both species investigated in our two companion studies, *Crassula helmsii* and *T. caerulescens*. Using more copper concentrations was impossible for practical reasons like measuring time at the FKM and beamtime at the synchrotron (almost one month already, which is longer than most other studies are allowed to take). We used a higher (but still sublethal!) copper concentration than Benzarti et al. (2008) for better comparability with *C. helmsii* and because otherwise the copper concentration in the *T. caerulescens* plants would have remained below the detection limit of EXAFS – already under our conditions we needed 3 days of synchrotron measuring time per sample. At half the concentration in the tissue it would have been about 12 days for the same signal/noise ratio, making the measurement of sufficient replicates impossible even when working with only one copper concentration as explained above. Further, we grew the plants for 4 months rather than 5 days in the study of Benzarti et al. (2008) because we wanted to investigate the long-term responses of the plant as these are more relevant than short-term toxicity for survival of the plant and for phytoremediation purposes. All chemicals except the EDDHA were analytical grade and purchased from Merck (Darmstadt, Germany; www.merck.de); Fe(III)-EDDHA was purchased from Duchefa Biochemie (Haarlem, The Netherlands; www.duchefa.com).

This study altogether lasted for nine years (2000-2009), and the experiments were carried out in two blocks with the development of a new fluorescence kinetic microscope (FKM) and data analysis in between. Altogether copper stress+acclimation was investigated in 7 experiments. In our first two experiments that included copper treatments of *T. caerulescens* (2000-2001, data analysis 2001-2003), we used 1.5 l vessels with 3-4 plants each and renewed the solutions manually every

4 d (i.e. renewal rate per plant about 85 ml.d^{-1}). In the second series (five experiments, 2004-2006, data analysis 2007-2009), we used 6 l vessels with 7 plants each, in which the solution was exchanged continuously (1700 ml.d^{-1} per pot, i.e. 250 ml.d^{-1} per plant) with the programmable 24-channel peristaltic pump "MCP Process" (Ismatec, Glattbrugg, Switzerland, www.ismatec.com). The solutions in the pots were constantly thoroughly mixed via a lab-built media injection system (Küpper et al. 2007a). The increased flow rate in the second experiment series was chosen to make sure that Cu uptake into the plants was not limited by the total amount available in the solution, but only by the concentration. While for Cd in *T. caerulescens* this yielded stronger stress compared to the earlier experiments (Küpper et al. 2007a) showing that Cd toxicity was limited by the total Cd per pot due to hyperaccumulation in the plants, for copper we did not observe a measurable difference between the high and low flow rate in the second vs. first experiment series.

All plants were grown with 14 h day length. In the first series of experiments, 24 °C/20 °C day/night temperature, and a constant irradiance of 60 µE (from a 1:1 mixture of "cool white" and Fluora® fluorescent tubes, OSRAM, München, Germany, www.osram.com) during the light period was applied. In the second series, 22°C/18°C day/night temperature was applied and a quasi-sinusoidal 3-step light cycle with about 40 µE in the morning and 120 µE at noon was achieved by full spectrum discharge lamps.

Despite the differences in growth conditions between the first and the second series of experiments, all trends of changes in photosynthetic parameters and growth were found in both series, so that all experiments will be analysed together in the results and discussion.

Chlorophyll fluorescence kinetic measurements

These were done with the "Fluorescence Kinetic Microscope" (FKM, developed by Küpper et al. 2000b; Küpper et al. 2007a) as described in the companion paper on *Crassula helmsii*, except for keeping the leaves in water-saturated air. To perform a measurement, a leaf was cut off and pressed by its upper side (for palisade mesophyll measurements) or lower side (for spongy mesophyll measurements) towards the glass window of the measuring chamber with a wet nylon grid or wet cellophane. The chamber was ventilated by a stream of water-saturated air (2 l.min^{-1}, 21°C). The construction and operation of the chamber is in principle described in Küpper et al. (2000b); the new version is shown Küpper et al. (2007a). All measurements were done on the mesophyll away from the veins.

Analysis of fluorescence kinetics.

The original data, i.e. two-dimensional records of chlorophyll fluorescence kinetics, were analysed using the FluorCam6 software from Photon Systems Instruments (Brno, Czech Republic) as described earlier (Küpper et al. 2000b; 2007a) and in our companion paper on *C. helmsii*. All parameters of fluorescence kinetics used in this study are explained in the list of abbreviations; a more detailed description on the technique as such and on all basic parameters can be found in the review of Maxwell and Johnson (2000).

Preparation of plant samples for EXAFS measurements

EXAFS samples were taken during the second series of experiments (see above). After about six months of plant growth, samples were taken from leaves of several developmental stages. To eliminate problems of element re-distribution during sample preparation, the collected tissues were shock-frozen in melting nitrogen slush. This slush was generated by pulling a strong vacuum on a container of well insulated liquid nitrogen – this forces the nitrogen into solid + gaseous state - and then releasing the vacuum, causing the nitrogen "snow" to melt. This rapid-freeze method prevents, in contrast to liquid nitrogen, the formation of a gas layer that slows down freezing of the sample. Afterwards, samples were immediately stored at –80 °C. To obtain representative datasets within the limited Synchrotron beamtime, aliquots of samples from all plants of the same metal treatment (see above) were mixed, so that each EXAFS sample represented the average of >10 plants. For the EXAFS measurements, the frozen-hydrated samples were ground to powder in a mortar and filled into EXAFS cuvettes; also all this was done at –80°C (dry ice cooling). Afterwards the cuvettes were sealed with Kapton[R] tape and stored in liquid nitrogen.

Preparation of copper complexes for spectroscopic measurements

A 5 mM solution of $CuSO_4$ in water was used as reference for the aquo complex. Histidine, citrate and glutathione complexes were made by adding 50 mM ligand to a 5 mM $CuSO_4$ solution in water. Our citrate and aquo complexes were acidic, adjusted to pH4, resembling the situation in plant vacuoles, which are the compartments where plants store organic acids, and which are main sinks for heavy metals in hyperaccumulators (Küpper et al. 1999, 2001). The histidine and glutathione complexes were prepared both at pH 4 and pH 7. Samples of Cu(II)-nicotianamine were prepared by potentiometric titration of a solution of 2.5 mM ligand and 2 mM $CuCl_2$ in water under the exclusion of dioxygen (nitrogen atmosphere). Aliquots of 0.1 M KOH were added stepwise to reach pH 11. Samples for UV/VIS and EPR spectroscopy were taken at pH 2, 3, 4, 5, 6, 7, 8, 9, 10 and 11; samples for EXAFS spectroscopy were taken at pH 4 and pH 7. Hereafter the solution was titrated back from pH 11 to pH 2 with 0.1 M HCl. For EXAFS and EPR spectroscopy, 10% (v/v) glycerol were added to all solutions of the model complexes to minimise the formation of ice crystals during freezing. Those samples were transferred into EXAFS cuvettes and frozen in liquid nitrogen, or into EPR tubes and frozen in supercooled (- 140 °C) isopentane.

EXAFS measurements and data analysis

Measurements were performed at the EMBL bending magnet beamline D2 (DESY, Hamburg, Germany) using a Si(111) double crystal monochromator, a focusing mirror and a 13-elements Ge solid-state fluorescence detector. All samples were mounted in a top-loading closed-cycle cryostat (modified Oxford instruments) and kept at about 30 K. The transmitted beam was used for energy calibration by means of the Bragg reflections of a static Si(220) crystal (Pettifer and Hermes, 1985). At least 500,000 counts above the Cu K absorption edge were accumulated for each measurement. Data reduction and average was done using EXPROG (Nolting et al. 1992) and KEMP (Korbas et al. 2006) packages of programs. Data analysis was performed using DL_Excurv (Tomic et al. 2005), which is a freeware version of Excurve (Binstead et al. 1992) under the flagship of CCP3 (www.ccp3.ac.uk/).

The k-range was chosen according to the statistical quality of the data at the end of the k-range, from 3 $Å^{-1}$ to a higher end at 8.5-13 $Å^{-1}$. In doubtful cases, the fits were performed in an extended and a shortened k-range, in order to test the significance of the low scattering contributions in the fit. Since N and O are indistinguishable by EXAFS due to their similar scattering phases, both elements are used indistinctively. But multiple scattering contributions, e.g. of the imidazole ring of histidine, may allow unambiguous assignment of ligand molecules (e.g. histidine, nicotianamine). Several models were attempted for the refinement, with different coordination numbers and mixtures of ligands in the first shell. The individual contributions of the potential ligands were floated during the refinement; in this way the possibility of several Cu species was tested. In the cases where scattering contributions were seen beyond the first shell, mixed first shells including histidine, nicotianamine and other low Z ligands (N/O) were tried. For explaining the contributions at long distances from the central atom, many models were tried, as described in the results. Comparison with a copper foil spectrum indicated that artefacts of contaminating scattered photons from the cryostat copper should have been minimal. Trying to refine such a contribution led only to very low percentages (usually <5% of total copper). In addition the intensity of the 5 Å peak was highest in samples with rather high copper content (i.e. the resistant specimens), while a contribution from cryostat reflections should have been strongest in samples low in internal copper. This was reproducibly found in 2 measuring years.

Histidine was modelled using the coordinates provided by the DL-Excurve release. The nicotianamine (NA) coordinates were taken from NICOAM01.cif (Miwa et al. 1999). The initial NA structure was taken from Nomoto et al. (1981). The Cu model compound of Nomoto el al. is 6-coordinated, with two longer Cu-ligand bonds. For the phase calculations of NA, a metal site with a bound nicotianamine was built using Spartan Student Edition v1.0.2 for MAC, with the ligand assumed to bind by the N on the plane of the ring and angles (Cu-N-C) of 121 degrees with each neighbouring carbon. In both cases, for histidine and nicotianamine rings, multiple scattering within the ring was included. Also binding from the carboxyl groups of nicotianamine was included.

In addition to the EXAFS refinement, like in our previous study on Cd and Zn ligands in *T. caerulescens* (Küpper et al. 2004) we used a fit with a linear combination of all measured model complexes (aqueous, malate, citrate, nicotianamine, histidine, proline, glutathione) as an independent way to determine the proportions of ligand types binding the copper in plant tissues. This fit is subsequently called "component analysis" (CA). The fitting of measured XAS sample data by a linear combination of measured model data has also successfully been used by Salt et al. (1999) for Zn in samples of *T. caerulescens*. In the present work the approach of Salt and co-workers was modified as described by Küpper et al. (2004).

Determination of metals in whole plant tissues

Frozen-hydrated plant samples (see above) were lyophilised, ground and subsequently digested with a mixture of 85% (v/v) concentrated HNO_3 and 15% (v/v) concentrated $HClO_4$ (Zhao et al. 1994). Concentrations of Cu in the digests were determined using atomic absorption spectrometry (AAS) in a GBC 932 AA spectrometer (GBC Scientific Equipment Pty Ltd., Dandenong, Australia).

UV/VIS spectroscopy and pigment analysis

UV/VIS spectra were measured with the double beam spectrometer Lambda 16 (Perkin Elmer, Wellesley, USA, las.perkinelmer.com) at 22±2 °C with a scanning speed of 240 $nm.min^{-1}$. An optical path length of 10 mm, 1 nm spectral bandwidth and 0.2 nm sampling interval was selected for all measurements. For pigment quantification, leaves were frozen in liquid nitrogen, then lyophilised, and finally extracted in 100% acetone. Chls and carotenoids were quantified according to Küpper et al. (2007c) as described in detail in our companion paper on *Crassula helmsii*.

Electron Paramagnetic Resonance spectroscopy

EPR spectra (perpendicular mode, X-band) were recorded on a Bruker Elexsys 500 with an ER 049 X microwave bridge (Bruker BioSpin GmbH, Rheinstetten, Germany; http://www.bruker-biospin.com). The system was equipped with an Oxford Instruments ESR 900 helium cryostat controlled by the ITC 503 temperature device. The modulation frequency was 100 kHz and the modulation amplitude was typically 0.1 mT. The measurements were performed with a Bruker 4122 SHQE cavity at ≈ 9.34 GHz. The microwave power was adjusted to obtain maximum signal intensity at the given temperature of 20K. Sample tubes were Suprasil quartz tubes 705-PQ-9.50 (Wilmad) with a $Ø_{out}$ of 4 mm, sample volume 250 µl. Samples were frozen in liquid nitrogen and kept at 77 K. Spectra were evaluated with the Xepr software (Bruker), and $CuSO_4$ in 2M $NaClO_4$/HCl, pH 1.5, served as the standard for quantitation.

ACKNOWLEDGEMENTS

We are very grateful to Ivan Šetlík for stimulating discussions about interpretation of the fluorescence kinetic parameters.

Table I. Plant growth and Cu concentrations in the plant samples. The values are means and standard errors of 3-10 plants.

Sample	Cu in the EXAFS samples / mg/kg dry weight (± se), n = number of plants used for EXAFS	Plant fresh weight (g ± se) after 8 weeks growth n = number of plants used for fresh weights	Plant fresh weight (g ± se) after 4 months growth n = number of plants used for fresh weights
Thlaspi caerulescens, 0.1 µM Cu^{2+}		5.31 (1.60) n = 4	54.7 (4.6) n=6
Thlaspi caerulescens, 10 µM Cu^{2+}, resistant individuals	42.7 (8.1) n = 5	-	6.33 (1.31) n=3
Thlaspi caerulescens, 10 µM Cu^{2+}, sensitive individuals	34.4 (3.8) n = 8	0.043 (0.041) n = 4	2.08 (0.17) n=10
Thlaspi fendleri, 0.1 µM Cu^{2+}	-	2.66 (1.06) n = 4	-
Thlaspi fendleri, 10 µM Cu^{2+}	-	0.04 (0.042) n = 4	-

- = not determined

Table II. Results of the refinement (= explanation with a theoretical model) of the EXAFS spectra of model compounds using the DL-Excurve program. The graphs of the fits for Cu (I/II)-glutathione, Cu(II)-histidine and Cu(II)-nicotianamine at pH 7 are shown in Fig. 3. The refinements of model compounds that were more relevant for *C. helmsii* (aqueous Cu(II), Cu(II)-citrate and Cu(II)-malate) are shown in the companion paper about copper metabolism in that species.

EF = Fermi energy, defines the threshold for the EXAFS spectra (Rehr and Albers, 2000). This value was refined for every sample.
FI = Fit index– increases with decreasing quality of the fit.
se = Mathematical standard errors of the refinement (two sigma level). The error of the EXAFS approach as such is higher, this is revealed by the differences between samples of the same type.
(1) = Fit index is higher due to the lower signal/noise ratio of the measured data.
$ mixture of Cu(I) and Cu(II) as obtained by the preparation method
# bound to histidine
*: Values were constrained to be identical for these contributions. Deviations from this simplification do not improve the model significantly.

Sample	Number / Type of ligands per shell (±se)	Distance (± se) [Å]	$2\sigma_i^2$ (± 2 sigma standard deviation) [Å^2]	EF (± se)	FI (number of variables)
Cu-Glutathione$^\$$	3.6 (±0.2) S	2.260(±0.002)	0.006(±0.001)	-11.5(±0.4)	0.142
	4.5 (±0.6) Cu	2.710(±0.005)	0.021(±0.002)		
Cu(II)-histidine	5.0 (±0.8) N#	2.000(±0.006)*	0.008(±0.001)*	-10.5(±0.6)	0.403
	0.4 (±0.8) O	2.000(±0.006)*	0.008(±0.001)*		
Cu(II)-nicotianamine pH 4	3.8 (±0.3) N/O	1.980(±0.003)	0.009(±0.001)	-9.8(±0.4)	0.251
	3.7 (±0.8) C	2.862(±0.012)	0.011(±0.004)*		
	5.0 (±1.3) C	3.403(±0.015)	0.011(±0.004)*		
Cu(II)-nicotianamine pH7	5.6 (±0.3) N/O	1.975(±0.003)	0.006 (±0.001)	-9.4(±0.5)	0.305
	4.1 (±0.8) C	2.837(±0.010)	0.006 (±0.003)*		
	4.7 (±1.2) C	3.394(±0.014)	0.006 (±0.003)*		
Cu(II)-proline pH7	5.2 (±0.3) N/O	1.957(±0.004)	0.008 (±0.001)	-12.5(±0.5)	0.285

Table III. Results of the refinement (= explanation with a theoretical model) of the EXAFS spectra of plant samples using the DL-Excurve program. To obtain representative datasets within the limited Synchrotron beamtime, aliquots of samples from all similarly aged leaves of the same plant were mixed. When available, leaves of several individuals from the same experiment with similar Cu-resistance were mixed as well.
\# = shown in Figure 5
EF = Fermi energy, defines the threshold for the EXAFS spectra (Rehr and Albers, 2000). FI = Fit index – increases with decreasing quality of the fit. se = Mathematical standard errors of the refinement (two sigma level). The error of the EXAFS approach as such is higher, this is revealed by the differences between samples of the same type.
*1, *2, *: Values were constrained to be identical for these contributions. Deviations from this simplification do not improve the model significantly. The amplitude reduction factor (Rehr and Albers, 2000) was set to 0.75. For nicotianamine (NA) the structural model obtained in the refinement of the model complex has been used, keeping the metal carbon distances constant and defining their Debye-Waller factors to 120% of the corresponding first shell value.

Sample incl. number of individual plants used to prepare it	Number of ligands ($\pm$se)	Distance ($\pm$ se) [Å]	$2\sigma_i^2$ ($\pm$ 2 sigma standard deviation) [Å^2]	EF ($\pm$ se) [eV]	FI · k^3
resistant individual, young leaves, 1st sample	1.6 ($\pm$0.4) S	2.279($\pm$0.015)	0.004 ($\pm$0.003) *1	-7.0 ($\pm$2.0)	1.710
	0.1 ($\pm$1.1) O	1.962($\pm$0.021)*	0.004 ($\pm$0.003) *1		
	2.2 ($\pm$1.7) NA	1.962($\pm$0.021)*	0.004 ($\pm$0.003) *1		
	2.5 ($\pm$1.7) P	2.825($\pm$0.023)	0.011 ($\pm$0.010)		
	1.4 ($\pm$0.7) Cu	3.388($\pm$0.017)	0.004 ($\pm$0.005) *2		
	0.8 ($\pm$0.8) Cu	4.333($\pm$0.046)	0.004 ($\pm$0.005) *2		
	6.1 ($\pm$3.3) Cu	5.207($\pm$0.016)	0.004 ($\pm$0.005) *2		
2 resistant individuals, young leaves, 2nd sample, \#	2.0 ($\pm$0.3) S	2.289($\pm$0.010)	0.011 ($\pm$0.002) *1	-6.9 ($\pm$1.3)	0.773
	0.1 ($\pm$1.5) O	1.932($\pm$0.022)*	0.011 ($\pm$0.002) *1		
	2.3 ($\pm$1.7) NA	1.932($\pm$0.022)*	0.011 ($\pm$0.002) *1		
	3.2 ($\pm$0.9) P	2.738($\pm$0.009)	0.010 ($\pm$0.004)		
	0.5 ($\pm$0.2) Cu	3.637($\pm$0.024)	0.004 ($\pm$0.002) *2		
	1.3 ($\pm$0.4) Cu	4.391($\pm$0.016)	0.004 ($\pm$0.002) *2		
	3.4 ($\pm$0.3) Cu	5.212($\pm$0.012)	0.004 ($\pm$0.002) *2		
2 resistant individuals, young leaves, 3rd sample,	1.4 ($\pm$0.3) S	2.286($\pm$0.011)	0.004 ($\pm$0.003) *1	-8.7 ($\pm$1.5)	1.213
	1.7 ($\pm$0.5) O	1.947($\pm$0.013)*	0.004 ($\pm$0.003) *1		
	0.2 ($\pm$0.6) NA	1.947($\pm$0.013)*	0.004 ($\pm$0.003) *1		
	2.0 ($\pm$1.0) P	2.848($\pm$0.016)	0.007 ($\pm$0.007)		

	1.0 (±0.6) Cu	3.365(±0.016)	0.004 (±0.005) [*2]		
	1.2 (±0.8) Cu	4.376(±0.023)	0.004 (±0.005) [*2]		
	4.1 (±2.0) Cu	5.200(±0.013)	0.004 (±0.005) [*2]		
resistant individual, mature leaves, 1st sample, #	1.1 (±0.2) S	2.307(±0.012)	0.009 (±0.002) [*1]	-12.3 (±0.9)	1.071
	0.1 (±0.3) O	1.945(±0.011)*	0.009 (±0.002) [*1]		
	2.5 (±0.4) NA	1.945(±0.011)*	0.009 (±0.002) [*1]		
	0.7 (±0.6) P	3.023(±0.020)	0.004 (±0.010)		
	0.3 (±0.3) Cu	3.668(±0.033)	0.004 (±0.006) [*2]		
	0.1 (±0.3) Cu	4.379(±0.151)	0.004 (±0.006) [*2]		
	2.5 (±1.7) Cu	5.233(±0.011)	0.004 (±0.006) [*2]		
4 resistant individuals, mature leaves, 2nd sample	2.1 (±0.4) S	2.245(±0.008)	0.006 (±0.003) [*1]	-7.2 (±1.4)	0.972
	0.1 (±1.8) O	1.912(±0.037)*	0.006 (±0.003) [*1]		
	1.6 (±2.1) NA	1.912(±0.037)*	0.006 (±0.003) [*1]		
	1.0 (±0.9) P	2.819(±0.019)	0.004 (±0.008)		
	0.6 (±0.3) Cu	3.705(±0.034)	0.004 (±0.002) [*2]		
	2.9 (±0.5) Cu	4.389(±0.011)	0.004 (±0.002) [*2]		
	3.7 (±1.0) Cu	5.216(±0.012)	0.004 (±0.002) [*2]		
sensitive individual, young leaves, 1st sample, #	1.4 (±0.2) S	2.275(±0.008)	0.006 (±0.002) [*1]	-9.5 (±0.9)	0.710
	1.9 (±0.8) O	1.933(±0.011)*	0.006 (±0.002) [*1]		
	0.6 (±0.9) NA	1.933(±0.011)*	0.006 (±0.002) [*1]		
	1.0 (±0.5) P	2.871(±0.017)	0.004 (±0.006)		
	0.7 (±0.6) Cu	3.719(±0.092)	0.011 (±0.008) [*2]		
	0.8 (±1.0) Cu	4.347(±0.045)	0.011 (±0.008) [*2]		
	3.7 (±2.5) Cu	5.229(±0.042)	0.011 (±0.008) [*2]		
2 sens. individuals, mature leaves, 2nd sample, #	1.1 (±0.2) S	2.255(±0.008)	0.006 (±0.002) [*1]	-8.6 (±1.3)	0.876
	0.1 (±0.3) O	1.907(±0.027)*	0.006 (±0.002) [*1]		
	2.5 (±0.4) NA	1.907(±0.027)*	0.006 (±0.002) [*1]		
	0.7 (±0.6) P	2.872(±0.043)	0.005 (±0.010)		
	0.3 (±0.3) Cu	3.306(±0.036)	0.009 (±0.008) [*2]		
	0.1 (±0.3) Cu	4.386(±0.020)	0.009 (±0.008) [*2]		
	2.5 (±1.7) Cu	5.238(±0.033)	0.009 (±0.008) [*2]		
5 sens. individuals, young-mature leaves, 3rd sample	0.4 (±0.2) S	2.333(±0.027)	0.007 (±0.002) [*1]	-9.2 (±0.8)	0.747
	0.1 (±0.7) O	1.958(±0.008)*	0.007 (±0.002) [*1]		
	3.6 (±0.8) NA	1.958(±0.008)*	0.007 (±0.002) [*1]		
	0.3 (±0.3) P	2.951(±0.047)	0.004 (±0.012)		

0.4 (±0.3) Cu	3.659(±0.026)	0.004 (±0.005) [*2]
0.1 (±0.3) Cu	4.366(±0.151)	0.004 (±0.005) [*2]
3.0 (±1.7) Cu	5.217(±0.010)	0.004 (±0.005) [*2]

3. General discussion

In this thesis, several aspects of hyperaccumulators dealing with accumulated and non-accumulated metals have been investigated. All studies concerned, directly or indirectly, the transport of metals in hyperaccumulators and dealed with the question how these plants manage to sequester the toxic metals to their final storage site. In the case of copper the question, how it is complexed to do as little harm to the plant as possible, has been asked.

Observation of cadmium acclimation using chlorophyll fluorescence kinetics

First, plants and their long-term reactions on cadmium were studied (Küpper et al. 2007, Chapter 2.1. in this thesis). During an ongoing project, *Thlaspi caerulescens* has been grown for 6 months under physiologically relevant conditions. This means that the plants were supplied with a medium with added amounts of metal as they could be experienced by plants in nature. Unfortunately, in many stress experiments, extremely high metal concentrations are used (examples: 120µM cadmium in a nutrient solution for Matricaria chamomilla (Kováčik et al. 2009), 100µM cadmium in a nutrient solution for Lycopersicon esculentum (Lopez-Millan et al. 2009). Metal concentrations too high make these studies useless for understanding how plants really deal with such stress in their natural habitat, like regions where mining has taken place. Also for developing better strategies for phytoremediation and phytomining, such studies are not of much use as it is important to understand how plants deal with the metal over their complete life-period. With our long-term approach, physiologically relevant metal concentrations and a sophisticated instrument, the Fluorescence Kinetic Microscope (FKM), we found that, after a period of stress, the plants acclimated to the cadmium supplied with the medium. During the acclimation process a heterogenity of accumulation appeared as some mesophyll cells took up more cadmium than others and thus showed stronger inhibition. Observing different photosynthetic parameters we found that Cd affects the photosynthetic light reactions more than the Calvin-Benson cycle and that Cd inhibits at least two different targets in/ around photosystem II. Further, acclimation in *T. caerulescens* is at least partially inducible and involves transient physiological heterogeneity as an emergency defence mechanism. Some of the mesophyll cells take up more metal than others until finally the greatest part of the cadmium is stored in the final storage site, the vacuole of large epidermal cells (Küpper et al. 1999).

Visualisation and quantitative analysis of Cd uptake into protoplasts

Thlaspi´s ability to take up and store such high amounts of metal without showing more symptoms of toxicity is certainly due to its ability to specifically pump the cadmium into its storage cells. We wanted to follow the way of the hyperaccumulated metal visually and find out the time limiting step in the process of sequestration (Leitenmaier and Küpper 2010, Chapter 2.2. in this thesis). To investigate this, again *Thlaspi caerulescens* was grown, this time in medium with a minimum of zinc and no cadmium. The leafs were digested, resulting in cells without cell wall, i.e. protoplasts. Not only mesophyll protoplasts were obtained, but also large epidermal storage cell protoplasts. This was particularly difficult and several earlier attempts to obtain them had failed (Cosio et al. 2004, 2005), as due to their large size they ruptured very easily. With the use of a fluorescent dye specific for cadmium and measurement in the FKM it was possible to follow the uptake of cadmium into the cytoplasm and finally into the vacuole. So far, only one study exists dealing with protoplasts and a cadmium specific fluorescent dye (Lindberg et al. 2004), but here wheat was investigated and only uptake into root and mesophyll cells has been studied. Our comparison of uptake rates into mesophyll and epidermal storage cells yielded the interesting result that large epidermal storage cells show significantly higher uptake rates compared to mesophyll and also normal sized epidermal cells. This is not very surprising, as these large storage cells are "dumping sites" for metal in *T. caerulescens*. As we´ve used protoplasts it is also clear, that these differences are not due to the transpiration stream but due to specific expression of metal transporters in the cytoplasmic membrane of storage cells.

Within the measured mesophyll cells, uptake rates varied. This finding fits very well with the conclusions of Küpper et al. 2007, where a heterogenity of cadmium distribution in mesophyll cells during cadmium acclimation was found. A new interesting finding was a bright signal in the cytoplasm shortly after the addition of cadmium to the medium. The signal stayed there for some time and then slowly translocated into the vacuole, indicating that the time limiting step in cadmium uptake into large storage cells must be the transport over the vacuolar membrane. Only little is known about vacuolar metal transporters in plants, but it has been suggested that CDFs like MTP1 (Desbrosses-Fonrouge et al. 2005, Küpper and Kochian, 2010) and the NRAMP family might play a role in metal transport over the vacuolar membrane (Oomen et al. 2008, Lanquar et al. 2010). Additionally, MTP1 has been found to be highly overexpressed in epidermal storage cells of *T. caerulescens* (Küpper and Kochian, 2010) suggesting a key role of MTP1 in sequestration of metal into to vacuole..

Insights into the function of TcHMA4, a Cd/Zn ATPase

Knowledge about metal transporters, especially ATPases for translocation over the cytoplasmic membrane of cells, has been increasing in recent years. But so far mainly small domains of proteins have been structurally resolved (Zimmermann et al. 2009, González-Guerrero, 2008) and biophysically characterised (Lübben et al. 2007). Due to the problems with overexpression of membrane proteins and the generally high content of cysteins in these ATPases, not a single structure of a holoenzyme is available and only few characterisation data of holoenzymes have been obtained so far (Parameswaran et al. 2007, Hung et al. 2007). In the work of Parameswaran et al. 2007 (Chapter 2.3. in this thesis), we successfully established a purification protocol for the P_{1B}-type ATPase TcHMA4 from roots of the Zn/Cd hyperaccumulator *T. caerulescens* using the natural overexpression of metal transporters in this plant (Pence et al. 2000) and the stretch of 8 histidines at the C-terminus (Papoyan and Kochian 2004). After purification, ATPase activity assays have been conducted showing first, that the protein has been purified in its native state and second, that it can be activated by cadmium and zinc. Continuing this project (Leitenmaier et al. 2011, Chapter 2.4. in this thesis), the purification process has been further optimised yielding protein less preloaded with metal. With this achievement, it was possible to obtain very interesting activity data including the activation with copper, cadmium and zinc. Similar experiments with an eukaryotic P_{1B}-type ATPase have only been shown by Hung et al. in 2007, where they investigated a human Cu-ATPase, MNK, which leads to the serious Menkes disease when mutated. Hung et al. were able to purify the protein in small amounts from insect (*Spodoptera frugiperda*, fall armyworm) cells. With their strategy, similar to ours, not to use heterologous expression in E.coli, they circumvented problems with misfolding of eukaryotic membrane proteins that often occur when overexpressed in bacterial cells. Using a specially built temperature gradient thermostat, we obtained activation data at 10 different temperatures which allowed us to calculate TcHMA4s energy of activation from Arrhenius plots. Unfortunately, no data for comparison from other P_{1B}-type ATPases are available in the literature and even for bacterial proteins of the same family, the search did not yield any results. Comparing our energy values with Ca-ATPases from fish, the data from fish seem to be somewhat higher, between 50 and 120kJ/mol ATP (Landeira-Fernandez et al. 2004), while another study observing the activity of an H^+-ATPase from protoplasts of *Commelina* (commonly called dayflower) yielded an E_A value of 8 kJ.mol^{-1} (Willmer et al. 1995), which is slightly lower than the values obtained for TcHMA4.

For analysing the metal binding sites in the protein in terms of primary ligands, we carried out EXAFS measurements on the cadmium and zinc K-edges. The purified protein from two independent isolations had been loaded with cadmium before concentration and during the concentration process all excess metal not bound to the protein has been washed out. This process was necessary for obtaining a cadmium signal without a high background signal, as the cadmium concentration in samples of large proteins are generally rather low (Lane et al. 2005), a fact that makes these samples difficult to measure. For cadmium, we measured two independent samples at two different beamlines (DESY, Hamburg, Germany and SSRL, Stanford, CA, USA) and both of them yielded a very similar result: both Fourier transformed data sets show a very clear peak at a distance of 2.5Å from the cadmium atom suggesting that cadmium in the protein is mainly bound by sulfur from cysteins. The measurement on the zinc edge shows something different. The Fourier transformed data yielded a peak at 2Å, which is a distance characteristic for nitrogen suggesting that zinc in TcHMA4 is mainly bound by histidines. As both cysteines and histidines are very abundant in the protein, both results make sense although the zinc measurement needs to be repeated as due to a lack of samples and beam time we were not able to measure a second sample yet.

New findings on how *Thlaspi caerulescens* deals with Cu

Low concentrations of metal in EXAFS samples are also one major topic of Chapter 2.6. (Mijovilovich et al. 2009). Here, the metal of interest was copper as, apart from some resistant individuals, *Thlaspi caerulescens* shows copper toxicity at, compared to cadmium, low copper concentrations. This is an important fact when using *Thlaspi* plants for phytoremediation of cadmium contaminated soils as these soils are, due to mining activities, very often not only polluted by cadmium but also by copper. Thus the abilities of *T. caerulescens* for phytoremediation of such soils can be limited (Walker and Bernal, 2004; Benzarti et al. 2008). In the study of Mijovilovich et al. we wanted to learn more about the detoxification of copper via ligands in *T. caerulescens* and the reactions of resistant and non resistant individuals upon copper treatment. In contrast to the hyperaccumulated metals cadmium and zinc, only little is known about the speciation of copper in *T. caerulescens*, as the concentrations in the tissues are very low and therefor difficult to detect. In the case of EXAFS it is additionally difficult to avoid artefacts during sample preparation, as during the freezing membranes might break, leading to unspecific binding of copper to the cell wall or other non-physiological ligands. Unfortunately, in most of the EXAFS studies done so far involving copper in plants, either the concentrations used during plant growth were far too high to be regarded as physiologically relevant (Shi et al. 2008) or artefacts during the freezing process might have occured (Sahi et al. 2007). There is only one study available that used copper concentrations as found in the environment and did not produce obvious artefacts (Polette et al. 2000). Here, a Cu(II)phytochelatin complex has been identified responsible for the transport of copper and another, unknown complex has been suggested to be involved in copper storage in the resistant indicator plant *Larrea tridentata*.

In our study we found that some individuals of *T. caerulescens* were somewhat resistent to copper, but also these plants did not reach a copper concentration to be regarded as a copper hyperaccumulator (>100ppm). Further, chlorophyll fluorescence kinetic measurements showed that copper treatment lead to a decrease in overall photosynthetic activity. In particular, the light saturation increased suggesting that the number of functional PSII reaction centers decreased faster than the number of functional antenna complexes. The EXAFS measurements showed a very high sulfur signal, most likely due to metallothioneins. This is in particular interesting as metallothioneins do not contribute very much to the binding of zinc in T. caerulescencs (Cobbett and Goldsbrough, 2002). Furthermore, Cu-Cu interactions were found and comparison with literature suggests that they might originate from Cu moolooite, a hydrated crystalline form of Cu oxalate (Michalowicz et al. 1979). As only very few studies dealed with this phenomenon so far, biomineralisation of copper in *T. caerulescens* is an interesting topic for future research. This study clearly showed that hyperaccumulator plants can distinguish very well between hyperaccumulated and non-accumulated metals, as seen by the main ligands: copper is mainly bound by sulfur while cadmium and zinc are both mainly bound by oxygen ligands (Küpper et al. 2004).

Topics for future research

Many open questions have been answered during the experiments for this thesis, but certainly also many questions have arisen and they might be interesting topics for future research.

As a protocol for isolation of large epidermal protoplast has been established in a time consuming process, it would be very interesting to use this knowledge now for comparison of differences in uptake rates between different cell types with a non-hyperaccumulating relative of *T. caerulescens*. This applies to root cells as well. In addition to cadmium, also the uptake of zinc could be investigated using a specific fluorescent dye.

Another organism could also be tested for isolation of probably more stable TcHMA4, for example the hyperaccumulator *Arabidopsis halleri*, which also accumulates zinc and cadmium. But as the many cysteines in the sequence of the protein seem to be responsible for the instability of the protein, the same problem is likely to occur with any other species. Isolation and purification under anoxic conditions might be the better attempt, as probably even the high concentrations of reductant used so far could be avoided. If the stability of the protein could be increased in any way, the way for further characterisation would be open: observation of metal transport with fluorescent dyes, inhibitions studies, CD spectroscopy and even crystallisation for structure determination would be interesting topics for the future.

4. References

Argüello JM; Eren E, González-Guerrero M (2007) The structure and funtion of heavy metal P1B-type ATPases. *Biometals* 20: 233-248.

Ascone I, Meyer-Klaucke W, Murphy L (2003) Experimental aspects of biological x-ray absorption spectroscopy. *J Synchrotron Radiat* 10: 16–22.

Assunção AGL, Costa Martins PDA, De Folter S, Vooijs R, Schat H, Aarts MGM. (2001) Elevated expression of metal transporter genes in three accessions of the metal hyperaccumulator *Thlaspi caerulescens. Plant, Cell & Environment* 24: 217–226.

Assunção AGL, Schat H, Aarts MGM (2003) *Thlaspi caerulescens*, an attractive model species to study heavy metal hyperaccumulation in plants. *New Phytologist* 159: 351–360.

Axelsen KB, Palmgren MG (2001) Inventory of the superfamily of P-type ion pumps in Arabidopsis. *Plant Physiol.* 126: 696–706.

Baker AJM (1981) Accumulators and excluders-strategies in the response of plants to heavy metals. *Journal of Plant Nutrition* 3: 643–654.

Baker AJM, McGrath SP, Sidoli CMD, Reeves RD (1994) The possibility of situ heavy metal decontamination of polluted soils using crops of metal-accumulating plants. *Resources, Conservation, and Recycling* 11: 41–49.

Banci L, Bertini I, Ciofi-Baffoni S, D'Onofrio M, Gonnelli L, Marhuenda-Egea FC, Ruiz-Dueñas FJ (2002) Solution structure of the N-terminal domain of a potential copper-translocating P-type ATPase from *Bacillus subtilis* in the apo and Cu(I) loaded states. *J. Mol. Biol.* 317: 415-429.

Baxter I, Tchieu J, Sussman MR, Bountry M, Palmgren MG, Gribskov M, Harper JF, Axelsen KB (2003) Genomic Comparison of P-Type ATPase Ion Pumps in Arabidopsis and Rice. *Plant Phys.* 132: 618-628.

Becher M, Talke IN, Krall L, Krämer U (2004) Cross-species microarray transcript profiling reveals high constitutive expression of metal homeostasis genes in shoots of the zinc hyperaccumulator *Arabidopsis halleri. Plant Journal* 37: 251–268.

Beneš I, Schreiber K, Ripperger H, Kircheiss A (1983) Metal complex formation by nicotianamine, a possible phytosiderophore. *Experientia* 39: 261–262.

Benzarti S, Mohri S, Ono Y (2008) Plant response to heavy metal toxicity: comparative study between the hyperaccumulator *Thlaspi caerulescens* (ecotype Ganges) and nonaccumulator plants: lettuce, radish, and alfalfa. *Environ Toxicol* 23: 607–616.

Bernard D, Roosens N, Czernic P, Lebrun M, Verbruggen N (2004) A Novel CPx-ATPase from the cadmium hyperaccumulator *Thlaspi caerulescens. FEBS Lett.* 569: 140–148.

Bert V, Meerts P, Saumitou-Laprade P, Salis P, Gruber W, Verbruggen N (2003) Genetic basis of Cd tolerance and hyperaccumulation in *Arabidopsis halleri*. *Plant and Soil* 249: 9–18.

Bhatia NP, Walsh KB, Orlic I, Siegele R, Ashwath N, Baker AJM (2004) Studies on spatial distribution of nickel in leaves and stems of the metal hyperaccumulator *Stackhousia tryonii* using nuclear microprobe (micro-PIXE) and EDXS techniques. *Functional Plant Biology* 31: 1061–1074.

Bidwell SD, Crawford SA, Woodrow IE, Summer-Knudsen J, Marshal AT (2004) Sub-cellular localization of Ni in the hyperaccumulator *Hybanthus floribundus* (Lindley) F. Muell. *Plant, Cell & Environment* 27: 705–716.

Binstead N, Strange RW, Hasnain SS (1992) Constrained and restrained refinement in EXAFS data analysis with curved wave theory. *Biochemistry* 31: 12117–12125.

Boyd RS, Davis MA, Wall MA, Balkwill K (2002) Nickel defends the South African hyperaccumulator *Senecio coronatus* (Asteraceae) against *Helix aspersa* (Mollusca: Pulmonidae). *Chemoecology* 12: 91–97.

Boyd RS, Martens SN (1994) Nickel hyperaccumulated by *Thlaspi montanum* var. *montanum* is acutely toxic to an insect herbivore. *Oikos* 70: 21–25.

Briskin DP, Poole RJ (1983) Characterization of a K^+-Stimulated Adenosine Triphosphatase Associated with the Plasma Membrane of Red Beet. *Plant Physiol.* 71: 350-355.

Broadhurst CL, Chaney RL, Angle JS, Erbe EF, Maugel TK (2004) Nickel localization and response to increasing Ni soil levels in leaves of the Ni hyperaccumulator *Alyssum murale*. *Plant and Soil* 265: 225–242.

Brooks RR, Lee J, Reeves RD, Jaffre T (1977) Detection of nickeliferous rocks by analysis of herbarium species of indicator plants. *Journal of Geochemical Exploration* 7: 49–57.

Brooks RR (1998) Geobotany and hyperaccumulators. In RR Brooks, ed, Plants That Hyperaccumulate Heavy Metals. CAB International, Wallingford, UK, pp 55–94

Buchauer MJ (1973) Contamination of soil and vegetation near a zinc smelter by zinc, cadmium, copper, and lead. *Environmental Science and Technology* 7: 131–135.

Bull PC, Cox DW (1994) Wilson disease and Menkes disease: new handles on heavy-metal transport. *TIG* 10: 246-252.

Byrd J, Berger RM, McMillin DR, Wright CF, Hamer D, Winge DR (1988) Characterization of the copper-thiolate cluster in yeast metallothionein and two truncated mutants. *J Biol Chem* 263: 6688–6694.

Calderone V, Dolderer B, Hartmann HJ, Echner H, Luchinat C, Del Bianco C, Mangani S, Weser U (2005) The crystal structure of yeast copper thionein: the solution of a long-lasting enigma. *Proc Natl Acad Sci USA* 102: 51–56.

Callahan DL, Baker AJM, Kolev SD, Wedd AG (2006) Metal ion ligands in hyperaccumulating plants. *J Biol Inorg Chem* 11: 2–12.

Cedeno-Maldonado A, Swader JA, Heath RL (1972) The cupric ion as an inhibitor of photosynthetic electron transport in isolated chloroplasts. *Plant Physiology* 50: 698–701.

Chaney RL, Angle JS, McIntosh MS, Reeves RD, Li YM, Brewer EP, Chen KY, Roseberg RJ, Perner H, Synkowski EC, Broadhurst CL, Wang S, Baker AJM (2005) Using hyperaccumulator plants to phytoextract Soil Ni and Cd. *Zeitschrift für Naturforschung* 60c: 190–198.

Chisholm JE, Jones GC, Purvis OW (1987) Hydrated copper oxalate, moolooite, in lichens. *Mineral Mag* 51: 715–718.

Ciscato M, Valcke R (1998) Chlorophyll fluorescence imaging of heavy metal treated plants. In: Garab G, ed. *Photosynthesis: mechanisms and effects*, Vol. IV. Dordrecht, the Netherlands: Kluwer Academic Publishers, 2661–2663.

Clarke RM, Williams IR (1986) Moolooite, a naturally occurring hydrated copper oxalate from Western Australia. *Mineral Mag* 50: 295–298.

Cobbett C, Goldsbrough P (2002) Phytochelatins and metallothioneins: roles in heavy metal detoxification and homeostasis. *Annual Review of Plant Biology* 53: 159–182.

Cobine PA, McKay RT, Zangger K, Dameron CT, Armitage IM (2004) Solution structure of Cu6 metallothionein from the fungus *Neurospora crassa*. *Eur J Biochem* 271: 4213–4221.

Cosio C, DeSantis L, Frey B, Diallo S, Keller C (2005) Distribution of cadmium in leaves of *Thlaspi caerulescens*. *Journal of Experimental Botany* 56: 765–775.

Cosio C, Martinoia E, Keller C (2004) Hyperaccumulation of cadmium and zinc in *Thlaspi caerulescens* and *Arabidopsis halleri* at the leaf cellular level. *Plant Physiology* 134: 716-725.

Coleman JOD, Randall R, Blake-Kalff MMA (1997) Detoxification of xenobiotics in plant cells by glutathione conjugation and vacuolar compartmentalization: a fluorescent assay using monochlorobimane. *Plant Cell and Environment* 20: 449-460.

Courbot M, Willems G, Motte P, Arvidsson S, Roosens N, Saumitou-Laprade P, Verbruggen N (2007) A Major Quantitative Trait Locus for Cadmium Tolerance in *Arabidopsis halleri* Colocalizes with *HMA4*, a Gene Encoding a Heavy Metal ATPase. *Plant Physiol.* 144: 1052-1065.

Cuypers A, Vangronsveld J, Clijsters H (2002) Peroxidases in roots and primary leaves of *Phaseolus vulgaris* copper and zinc phytotoxicity: a comparison. *Journal of Plant Physiology* 159: 869–876.

Desbrosses-Fonrouge AG, Voigt K, Schroder A, Arrivault S, Thomine S, Krämer U (2005) *Arabidopsis thaliana* MTP1 is a Zn transporter in the vacuolar membrane which mediates Zn detoxification and drives leaf Zn accumulation. *FEBS Letters* 579: 4165-4174.

Ebbs S, Lau I, Ahner B, Kochian LV (2002) Phytochelatin synthesis is not responsible for Cd tolerance in the Zn/Cd hyperaccumulator *Thlaspi caerulescens* (J. & C. Presl). *Planta* 214: 635–640.

Ebbs S, Lau I, Ahner B, Kochian LV (2002) Phytochelatin synthesis is not responsible for Cd tolerance in the Zn/Cd hyperaccumulator *Thlaspi caerulescens* (J.&C. Presl). *Planta* 214: 635-640.

Elbaz B, Shoshani-Knaani N, David-Assael O, Mizrachy-Dagri T, Mizrahi K, Saul H, Brook E, Berezin I, Shaul O (2006) High expression in leaves of the zinc hyperaccumulator *Arabidopsis halleri* of AhMHX, a homolog of an *Arabidopsis thaliana* vacuolar metal/proton exchanger. *Plant Cell and Environment* 29: 1179-90.

Faller P, Kienzler K, Krieger-Liszkay A (2005) Mechanism of Cd^{2+} toxicity: Cd^{2+} inhibits photoactivation of Photosystem II by competitive binding to the essential Ca^{2+} site. *Biochimica Biophysia Acta* 1706: 158–164.

Fergusson JE, Hayes RW, Yong TS, Thiew SH (1980) Heavy metal pollution by traffic in Christchurch, New Zealand: lead and cadmium content of dust, soil and plant samples. *New Zealand Journal of Science* 23: 293–310.

Ferimazova N, Küpper H, Nedbal L, Trtílek M (2002) New insights into photosynthetic oscillations revealed by two-dimensional microscopic measurements of chlorophyll fluorescence kinetics in intact leaves and isolated protoplasts. *Photochemistry and Photobiology* 76: 501–508.

Fomina M, Hillier S, Charnock JM, Melville K, Alexander IJ, Gadd GM (2005) Role of oxalic acid overexcretion in transformations of toxic metal minerals by *Beauveria caledonica*. *Appl Environ Microbiol* 71: 371–381.

Franck F, Dewez D, Popovic R (2005) Changes in the room-temperature emission spectrum of Chlorophyll during fast and slow phases of the Kautsky effect in intact leaves. *Photochemistry and Photobiology* 81: 431–443.

Franck F, Juneau P, Popovic R (2002) Resolution of the photosystem I and photosystem II contributions to chlorophyll fluorescence of intact leaves at room temperature. *Biochimica et Biophysica Acta: Bioenergetics* 1556: 239–246.

Frey B, Keller C, Zierold K, Schulin R (2000) Distribution of Zn in functionally different epidermal cells of the hyperaccumulator *Thlaspi caerulescens*. *Plant, Cell & Environment* 23: 675–687.

Gardea-Torresdey JL, Arteaga S, Tiemann KJ, Chianelli R, Pingitore N, Mackay W (2001) Absorption of copper(II) by creosote bush (*Larrea tridentata*): use of atomic and x-ray absorption spectroscopy. *Environ Toxicol Chem* 20: 2572–2579.

Genty B, Briantais J, Baker NR (1989) The relationship between the quantum yield of photosynthetic electron-transport and quenching of chlorophyll fluorescence. *Biochimica et Biophysica Acta* 990: 87–92.

George GN, Byrd J, Winge DR (1988) X-ray absorption studies of yeast copper metallothionein. *J Biol Chem* 263: 8199–8203.

González-Guerrero M, Eren E, Rawat S, Stemmler TL, Argüello JM (2008) Structure of the Two Transmembrane Cu^+ Transport Sites of the Cu+-ATPases. *J. Biol. Chem.* 283: 29753- 29759.

Guerinot ML (2000) The ZIP family of metal transporters. *Biochimica et Biophisica Acta* 1465: 190-198.

Ha SW, Korbas M, Klepsch M, Meyer-Klaucke W, Meyer O, Svetlitchnyi V (2007) Interaction of potassium cyanide with the [Ni-4Fe-5S] active site cluster of CO dehydrogenase from *Carboxydothermus hydrogenoformans*. *J Biol Chem* 282: 10639–10646.

Hall JL, Williams LE (2003) Transition metal transporters in plants. *J. Exp. Bot.* 54: 2601–2613.

Hanikenne M, Talke IN, Haydon MJ, Lanz C, Nolte A, Motte P, Kroymann J, Weigel D, Krämer U (2008) Evolution of metal hyperaccumulation required cis-regulatory changes and triplication of HMA4. *Nature* 453: 391-396.

Hanson B, Garifullina GF, Lindblom SD, Wangeline A, Ackley A, Kramer K, Norton AP, Lawrence CB, Pilon-Smits EAH (2003). Selenium accumulation protects *Brassica juncea* from invertebrate herbivory and fungal infection. *New Phytologist* 159: 461–469.

Haydon MJ, Cobbett CS (2007) A novel major facilitator superfamily protein at the tonoplast influences zinc tolerance and accumulation in *Arabidopsis*. *Plant Physiology* 143: 1705-1719.

Hemelrijk PW, Kwa SLS, van Grondelle R, Dekker JP (1992). Spectroscopic properties of LHC-II, the main light-harvesting chlorophyll a/b protein complex from chloroplast membranes. *Biochimica Biophysica Acta* 1098: 159–166.

Hollenstein K, Comellas-Bigler M, Bevers LE, Feiters MC, Meyer-Klaucke W, Hagedoorn PL, Locher KP (2009) Distorted octahedral coordination of tungstate in a subfamily of specific binding proteins. *J Biol Inorg Chem* 14: 663–672.

Hung YH, Layton MJ, Voskoboinik I, Mercer JFB, Camakaris J (2007) Purification and membrane reconstitution of catalytically active Menkes copper-transporting P-type ATPase (MNK; ATP7A). *Biochem. J.* 401: 569-579.

Hussain D, Haydon MJ, Wang Y, Wong E, Sherson SM, Young J, Camakaris J, Harper JF, Cobbett CS (2004) P-Type ATPase Heavy Metal Transporters with Roles in Essential Zinc Homeostasis in Arabidopsis. *Plant Cell* 16:1327-1339.

Irtelli B, Petrucci WA, Navarri-Izzo F (2009) Nicotianamine and histidine/ proline are, respectively, the most important copper chelators in xylem sap of *Brassica carinata* under conditions of copper deficiency and excess. *J Exp Bot* 60: 269–277.

Jhee EM, Boyd RS, Eubanks MD (2005) Nickel hyperaccumulation as an elemental defense of *Streptanthus polygaloides* (Brassicaceae): influence of herbivore feeding mode. *New Phytologist* 168: 331–343.

Jiménez-Ambriz G, Petit C, Bourrié I, Dubois S, Olivieri I, Ronce O (2007) Life history variation in the heavy metal tolerant plant *Thlaspi caerulescens* growing in a network of contaminated and noncontaminated sites in Southern France: role of gene flow, selection and phenotypic plasticity. *New Phytol.* 173: 199-215.

Joshi MK, Mohanty P (2004) Chlorophyll a fluorescence as a probe of heavy metal ion toxicity in plants. In: Papageorgiou GC, Govindjee, eds. *Chlorophyll a fluorescence: a signature of photosynthesis*. Dordrecht, The Netherlands: Springer, 637–661.

Koch M, Al-Shehbaz IA (2004) Taxonomic and phylogenetic evaluation of the American '*Thlaspi*' species: identity and relationship to the Eurasian genus *Noccaea* (Brassicaceae). *Systematic Botany* 29: 375–384.

Korbas M, Marsa DF, Meyer-Klaucke W (2006) KEMP: a program script for automated biological x-ray absorption spectroscopy data reduction. *Rev Sci Instrum* 77: 063105.

Kováčik J, Bořivoj K, Hedbavny J, Štork F, Bačkor M (2009) Comparison of cadmium and copper effect on phenolic metabolism, mineral nutrients and stress-related parameters in *Matricaria chamomilla* plants. *Plant Soil* 320: 231-242.

Krämer U, Cotterhowells JD, Charnock JM, Baker AJM, Smith JAC (1996) Free histidine as a metal chelator in plants that accumulate nickel. *Nature* 379: 635–638.

Krupa Z (1999) Cadmium against higher-plant photosynthesis – A variety of effects and where they possibly come from. *Zeitschrift für Naturforschung C* 54: 723–729.

Krupa Z, Öquist G, Huner NPA (1993) The effects of cadmium on photosynthesis of *Phaseolus vulgaris* – a fluorescence analysis. *Physiologia Plantarum* 88: 626–630.

Küpper H, Seibert S, Aravind P (2007c) A fast, sensitive and inexpensive alternative to analytical pigment HPLC: quantification of chlorophylls and carotenoids in crude extracts by fitting with Gauss-peak-spectra. *Anal Chem* 79: 7611–7627.

Küpper H, Seib LO, Sivaguru M, Kochian LV (2007b) A method for cellular localisation of gene expression via quantitative in situ hybridisation in plants. *Plant J* 50: 159–187.

Küpper H, Götz B, Mijovilovich A, Küpper FC, Meyer-Klaucke W (2009) Complexation and toxicity of copper in higher plants. I. Characterization of copper accumulation, speciation, and toxicity in *Crassula helmsii* as a new copper accumulator. *Plant Physiol* 151: 702–714.

Küpper H, Kroneck PMH. (2005) Heavy metal uptake by plants and cyanobacteria. In: Sigel A, Sigel H, Sigel RKO, eds. *Metal ions in biological systems*. New York, NY, USA: Marcel Dekker Inc, 97–142.

Küpper H, Küpper F, Spiller M (1996) Environmental relevance of heavy metal substituted chlorophylls using the example of water plants. *Journal of Experimental Botany* 47: 259–266.

Küpper H, Küpper F, Spiller M (1998) In situ detection of heavy metal substituted chlorophylls in water plants. *Photosynthesis Research* 58: 125–133.

Küpper H, Küpper FC, Spiller M (2006) [Heavy metal]-chlorophylls formed in vivo during heavy metal stress and degradation products formed during digestion, extraction and storage of plant material. In: Grimm B, Porra R, Rüdiger W, Scheer H, eds *Chlorophylls and bacteriochlorophylls: biochemistry, biophysics, functions and applications*. Advances in Photosynthesis and Respiration, Vol. 25. Dordrecht, the Netherlands: Kluwer Academic Publishers, 67–77.

Küpper H, Lombi E, Zhao FJ, McGrath SP (2000a) Cellular compartmentation of cadmium and zinc in relation to other elements in the hyperaccumulator *Arabidopsis halleri*. *Planta* 212: 75–84.

Küpper H, Lombi E, Zhao FJ, Wieshammer G, McGrath SP (2001) Cellular compartmentation of nickel in the hyperaccumulators *Alyssum lesbiacum*, *Alyssum bertolonii* and *Thlaspi goesingense*. *Journal of Experimental Botany* 52: 2291–2300.

Küpper H, Mijovilovich A, Meyer-Klaucke W, Kroneck PMH (2004) Tissue- and age-dependent differences in the complexation of cadmium and zinc in the Cd/Zn hyperaccumulator *Thlaspi caerulescens* (Ganges ecotype) revealed by X-ray absorption spectroscopy. *Plant Physiology* 134: 748–757.

Küpper H, Parameswaran A, Leitenmaier B, Trtílek M, Sétlík I (2007) Cadmium-induced inhibiton of photosynthesis and long-term acclimation to cadmium stress in the hyperaccumulator *Thlaspi caerulescens*. *New Phytol*. 175: 655–674.

Küpper H, Setlík I, Spiller M, Küpper FC, Prásil O (2002) Heavy metal-induced inhibition of photosynthesis: targets of in vivo heavy metal chlorophyll formation. *Journal of Phycology* 38: 429–441.

Küpper H, Setlík I, Trtílek M, Nedbal L (2000b) A microscope for two-dimensional measurements of *vivo* chlorophyll fluorescence kinetics using pulsed measuring light, continuous actinic light and saturating flashes. *Photosynthetica* 38: 553–570.

Küpper H, Spiller M, Küpper F (2000c) Photometric method for the quantification of chlorophylls and their derivatives in complex mixtures: fitting with gauss-peak-spectra. *Analytical Biochemistry* 286: 247–256.

Küpper H, Zhao FJ, McGrath SP (1999) Cellular compartmentation of zinc in leaves of the hyperaccumulator *Thlaspi caerulescens*. *Plant Physiology* 119: 305–311.

Küpper H. & Kochian LV (2010) Transcriptional regulation of metal transport genes and mineral nutrition during acclimation to cadmium and zinc in the Cd/Zn hyperaccumulator, *Thlaspi caerulescens* (Ganges population). *New Phytologist* 185: 114-129.

Küpper H, Šetlík I, Seibert S, Prášil O, Šetlikova E, Strittmatter M, Levitan O, Lohscheider J, Adamska I, Berman-Frank I (2008) Iron limitation in the marine cyanobacterium *Trichodesmium* reveals new insights into regulation of photosynthesis and nitrogen fixation. *New Phytologist* 179: 784-798.

Küpper H, Kroneck PMH (2007) Nickel in the environment and its role in the metabolism of plants and cyanobacteria, in: A. Sigel, H. Sigel, R.K.O. Sigel (Eds.), *Metal Ions in Life Sciences*, vol. 2, John Wiley and Sons, Ltd., pp. 31–62 (Chapter 2).

Laemmli UK (1970) Cleavage of structural proteins during the assembly of the head of bacteriophage T4. *Nature* 227: 680–685.

Lagerwerff JV, Specht AW (1970) Contamination of roadside soil and vegetation with cadmium, nickel, lead, and zinc. *Environmental Science and Technology* 4: 583–586.

Lanaras T, Moustakas M, Symeonidis L, Diamantoglou S, Karataglis S (1993) Plant metal content, growth responses and some photosynthetic measurements on field-cultivated wheat growing on ore bodies enriched in Cu. *Physiol Plant* 88: 307–314.

Landeira-Fernandez AM, Morrissette JM, Blank JM, Block BA (2004) Temperature dependence of the Ca^{2+}-ATPase (SERCA2) in the ventricles of tuna and mackerel. *Am. J. Physiol. Regul. Integ.r Comp. Physio.l* 286: 398–404.

Lane TW, Morel FMM (2000) A biological function for cadmium in marine diatoms. *Proceedings of National Academy of Science USA* 97: 4627–4631.

Lane TW, Saito MA, George GN, Pickering IJ, Prince RC, Morel FMM (2005) A cadmium enzyme from a marine diatom. *Nature* 435: 42.

Lanquar V, Schnell Ramos M, Lelièvre F, Barbier-Brygoo H, Krieger-Liszkay A, Krämer U, Thomine S (2010) Export of vacuolar manganese by AtNRAMP3 and AtNRAMP4 is required for optimal photosynthesis and growth under manganese deficiency. *Plant Physiol*. 152:1986-1999.

Lasat MM, Baker AJM, Kochian LV (1996) Physiological characterization of root Zn^{2+} absorption and translocation to shoots in Zn hyperaccumulator and nonaccumulator species of *Thlaspi*. *Plant Physiology* 112: 1715–1722.

Lasat MM, Baker AJM, Kochian LV (1998) Altered Zn compartmentation in the root symplasm and stimulated Zn absorption into the leaf as mechanisms involved in Zn hyperaccumulation in *Thlaspi caerulescens*. *Plant Physiology* 118: 875–883.

Leitenmaier B, Küpper H (2010) Cadmium uptake and sequestration kinetics in individual leaf cell protoplasts of the Cd/Zn hyperaccumulator *Thlaspi caerulescens*. Accepted for publication in *Plant Cell Envi*.

Liao MT, Hedley MJ, Woolley DJ, Brooks RR, Nichols MA (2000) Copper uptake and translocation in chicory (*Cichorium intybus L. cv Grasslands Puna*) and tomato (*Lycopersicon esculentum Mill. cv Rondy*) plants grown in NFT system. II. The role of nicotianamine and histidine in xylem sap copper transport. *Plant Soil* 223: 243–252.

Lindberg S, Landberg T, Greger M (2004) A new method to detect cadmium uptake in protoplasts. *Planta* 219: 526-532.

Lombi E, Zhao FJ, Dunham SJ, McGrath SP (2000) Cadmium accumulation in populations of *Thlaspi caerulescens* and *Thlaspi goesingense*. *New Phytologist* 145: 11–20.

López-Millán AF, Sagardoy R, Solanas M, Abadía A, Abadía J (2009) Cadmium toxicity in tomato (*Lycopersicon esculentum*) plants grown in hydroponics. *Env. Exp. Bot.* 65: 37-385.

Lübben M, Güldenhaupt J, Zoltner M, Deigweiher K, Haebel P, Urbanke C, Scheidig AJ (2007) Sulfate Acts as Phosphate Analog on the Monomeric Catalytic Fragment of the CPx-ATPase CopB from *Sulfolobus solfataricus*. *J. Mol. Biol.* 369: 368-385.

Lu L, Tian S, Yang X, Wang X, Brown P, Li I, He Z (2008) Enhanced root-to-shoot translocation of cadmium in the hyperaccumulating ecotype of Sedum alfredii. *Journal of Experimental Botany* 59: 3203-3213.

Ma JF, Ueno D, Zhao FJ, McGrath SP (2005) Subcellular localisation of Cd and Zn in the leaves of a Cd-hyperaccumulating ecotype of *Thlaspi caerulescens*. *Planta* 220: 731-736.

Macnair MR, Bert V, Huitson SB, SaumitouLaprade P, Petit D (1999) Zinc tolerance and hyperaccumulation are genetically independent characters. *Proceedings of the Royal Society of London, Series B* 266: 2175–2179.

Mana-Capelli S, Mandal AK, Argüello JM (2003) Archaeoglobus fulgidus CopB Is a Thermophilic Cu^{2+}-ATPase. *J. Biol. Chem.* 278: 40534-40541.

Marques L, Cossegal M, Bodin S, Czernic P, Lebrun M (2004) Heavy metal specificity of cellular tolerance in two hyperaccumulating plants, *Arabidopsis halleri* and *Thlaspi caerulescens*. *New Phytologist* 164: 289-295.

Martens SN, Boyd RS (1994) The ecological significance of nickel hyperaccumulation: a plant chemical defense. *Oecologia* 98: 379–384.

Maxted AP, Black CR, West HM, Crout NMJ; McGrath SP, Young SD (2007) Phytoextraction of cadmium and zinc from arable soils amended with sewage sludge using *Thlaspi caerulescens*: Development of a predictive model. *Environmental Pollution* 150: 363-372.

Maxwell K, Johnson GN (2000) Chlorophyll fluorescence: a practical guide.

Journal of Experimental Botany 51: 659–668.

McBride MB, Barrett KA, Martinez CE (2005) Zinc and cadmium distribution and leaching in a metalliferous peat. *Water Air and Soil Pollution* 171: 67–80.

McBride MB, Richards BK, Steenhuis T, Russo JJ, Sauvé S (1997) Mobility and solubility of toxic metals and nutrients in soil fifteen years after sewage sludge application. *Soil Science* 162: 487–500.

McGrath SP, Lombi E, Gray CW, Caille N, Dunham SJ, Zhao FJ (2006) Field evaluation of Cd and Zn phytoextraction potential by the hyperaccumulators Thlaspi caerulescens and Arabidopsis halleri. *Environmental Pollution* 141: 115-125.

McGrath SP, Zhao FJ (2003) Phytoextraction of metals and metalloids from contaminated soils. *Current Opinion in Biotechnology* 14: 277–282.

Michalowicz A, Girerd JJ, Goulon J (1979) EXAFS determination of the copper oxalate structure: relation between structure and magnetic properties. *Inorg Chem* 18: 3004–3010.

Mijovilovich A, Leitenmaier B, Meyer-Klaucke W, Kroneck PMH, Götz B, Küpper H (2009) Complexation and toxicity of copper in higher plants : II. Different mechanisms for Cu vs. Cd detoxification in the Cu sensitive Cd/Zn hyperaccumulator *Thlaspi caerulescens* (Ganges ecotype). *Plant Physiol.* 151: 715-731.

Mitani T, Ogawa M (1998) Cadmium leaching by acid rain from cadmiumenriched activated sludge applied in soil. *Journal of Environmental Science and Health* A33: 1569–1581.

Miwa Y, Mizuno T, Tsuchida K, Taga T, Iwata Y (1999) Experimental charge density and electrostatic potential in nicotinamide. *Acta Crystallogr B* 55: 78–84.

Morel M, Crouzet J, Gravot A, Auroy P, Leonhardt N, Vavasseur A & Richaud P (2009) AtHMA3, a P1B-ATPase allowing Cd/Zn/Co/Pb vacuolar storage in *Arabidopsis*. *Plant Physiol* 149: 894-904.

Nedbal L, Soukupová J, Kaftan D, Whitmarsh J, Trtílek M (2000) Kinetic imaging of chlorophyll fluorescence using modulated light. *Photosynthesis Research* 66: 3–12.

Nedbal L, Whitmarsh J (2004) Chlorophyll fluorescence imaging of leaves and fruits. In: *Chlorophyll a fluorescence – a signature of photosynthesis*. Advances in Photosynthesis and Respiration, Vol. 19. Dordrecht, the Netherlands, Springer: 389–407.

Nomoto K, Mino Y, Ishida T, Yoshioka H, Ota N, Inoue M, Takagi S, Takemoto T (1981) X-ray crystal-structure of the copper(II) complex of mugineic acid, a naturally-occurring metal chelator of graminaceous plants. *J Chem Soc Chem Commun* 7: 338–339.

Oomen RJFJ, Wu J, Lelièvre F, Blanchet S, Richaud P, Barbier-Brygoo H, Aarts MGM, Thomine S (2009) Functional characterization of NRAMP3 and NRAMP4 from the metal hyperaccumulator *Thlaspi caerulescens*. *New Phytol.* 181: 637-650.

Ouzounidou G, Moustakas M, Strasser RJ (1997). Sites of action of copper in the photosynthetic apparatus of maize leaves: kinetic analysis of chlorophyll fluorescence, oxygen evolution, absorption changes and thermal dissipation as monitored by photoacoustic signals. *Australian Journal of Plant Physiology* 24: 81–90.

Oxborough K (2004) Using chlorophyll a fluorescence imaging to monitor photosynthetic performance. In: Papageorgiou GC, Govindjee, eds. *Chlorophyll a fluorescence: a signature of photosynthesis*. Dordrecht, the Netherlands, Springer: 409–428.

Palmgren MG, Harper JF (1999) Pumping with plant P-Type ATPases. *J. Exp. Bot.* 50: 883–893.

Papageorgiou GC, Govindjee (2004) *Chlorophyll a fluorescence: a signature of photosynthesis*. Dordrecht, the Netherlands: Springer.

Papoyan A, Kochian LV (2004) Identification of *Thlaspi caerulescens* genes that may be involved in heavy metal hyperaccumulation and tolerance. Characterization of a novel heavy metal transporting ATPase. *Plant Physiology* 136: 3814–3823.

Parameswaran A, Leitenmaier B, Yang M, Welte W, Kroneck PMH, Lutz G, Papoyan A, Kochian LV, Küpper H (2007) A native Zn/Cd transporting P1B type ATPase protein from natural overexpression in a Zn/Cd hyperaccumulator plant. *Biochem Biophys Res Commun* 364: 51–56

Peer WA, Mahmoudian M, Freeman JL, Lahner B, Richards EL, Reeves RD, Murphy AS, Salt DE (2006) Assessment of plants from the Brassicaceae family as genetic models for the study of nickel and zinc hyperaccumulation. *New Phytologist* 172: 248–260.

Peer WA, Mamoudian M, Lahner B, Reeves RD, Murphy AS, Salt DE (2003) Development of a model plant to study the molecular genetics of metal hyperaccumulation. Part I: Germplasm analysis of 20 Brassicaceae accessions from Austria, France, Turkey, and USA. *New Phytologist* 159: 421–430.

Pence NS, Larsen PB, Ebbs SD, Letham DLD, Lasat MM, Garvin DF, Eide D, Kochian LV (2000) The molecular physiology of heavy metal transport in the Zn/Cd hyperaccumulator *Thlaspi caerulescens*. *Proceedings of National Academy of Sciences, USA* 97: 4956–4960.

Pettifer RF, Hermes C (1985) Absolute energy calibration of x-ray radiation from synchrotron sources. *J Appl Cryst* 18: 404–412

Pich A, Scholz I (1996) Translocation of copper and other micronutrients in tomato plants (*Lycopersicon esculentum* Mill.): nicotianaminestimulated copper transport in the xylem. *J Exp Bot* 47: 41–47.

Pich A, Scholz G, Stephan UW (1994) Iron-dependent changes of heavy metals, nicotianamine, and citrate in different plant organs and in the xylem of two tomato genotypes: nicotianamine as possible copper translocator. *Plant Soil* 165: 189–196.

Pilon M, Abdel-Ghany SE, Cohu CM, Gogolin KA, Ye H (2006) Copper cofactor delivery in plant cells. *Curr Opin Plant Biol* 9: 256–263.

Polette LA, Gardea-Torresdey JL, Chianelli RR, George GN, Pickering IJ, Arenas J (2000) XAS and microscopy studies of the uptake and biotransformation of copper in *Larrea tridentata* (creosote bush). *Microchem J* 65: 227–236.

Pollard AJ, Powell KD, Harper FA, Smith JAC (2002) The genetic basis of metal hyperaccumulation in plants. *Critical Reviews in Plant Sciences* 21: 539–566.

Porta P, Morpurgo S, Pettiti I (1996) Characterization by x-ray absorption, x-ray powder diffraction, and magnetic susceptibility of Cu-Zn-Co-Al containing hydroxycarbonates, oxycarbonates, oxides, and their products of reduction. *J Solid State Chem* 121: 372–378.

Poschenrieder C, Tolrà R, Barceló J (2006) Can metals defend plants against biotic stress? *Trends in Plant Science* 11: 288–293.

Prasad MNV, Hagemeyer J (1999) *Heavy metal stress in plants: from molecules to ecosystems*. Berlin, Germany: Springer.

Psaras GK, Constantinidis TH, Cotsopoulos B, Manetas Y (2000) Relative abundance of nickel in the leaf epidermis of eight hyperaccumulators: evidence that the metal is excluded from both guard cells and trichomes. *Annals of Botany* 86: 73-88.

Purvis OW, Pawlik-Skowronska B, Cressey G, Jones GC, Kearsley A, Spratt J (2008) Mineral phases and element composition of the copper hyperaccumulator lichen *Lecanora polytropa*. *Mineral Mag* 72: 607–616.

Raskin I, Smith RD, Salt DE (1997) Phytoremediation of metals: using plants to remove pollutants from the environment. *Curr. Opin. Biotechnol.* 8:221–226.

Rehr JJ, Albers RC (2000) Theoretical approaches to x-ray absorption fine structure. *Rev Mod Phys* 72: 621–654.

Rohácek K (2002) Chlorophyll fluorescence parameters: the definitions, photosynthetic meaning, and mutual relationships. *Photosynthetica* 40: 13–29.

Roosens NH, Leplae R, Bernard C, Verbruggen N (2005) Variations in plant metallothioneins: the heavy metal hyperaccumulator *Thlaspi caerulescens* as a study case. *Planta* 222: 716-729.

Roosens NH, Bernard C, Leplae R, Verbruggen N (2004) Evidence for copper homeostasis function of metallothionein (MT3) in the hyperaccumulator *Thlaspi caerulescens*. *FEBS Lett* 577: 9–16.

Ruban AV, Horton P (1994) Spectroscopy of non-photochemical and photochemical quenching of chlorophyll fluorescence in leaves; evidence for a role of the light-harvesting complex of photosystem II in the regulation of energy dissipation. *Photosynthesis Research* 40: 181–190.

Sachs J (1865) *Handbuch der Experimental-Physiologie der Pflanzen*. Leipzig, Germany: Verlag von Wilhelm Engelmann. pp. 153-154: §47.

Sagner S, Kneer R, Wanner G, Cosson JP, Deus-Neumann B, Zenk MH (1998) Hyperaccumulation, complexation and distribution of nickel in *Sebertia acuminata*. *Phytochemistry* 47: 339–347.

Sahi SV, Israra M, Srivastava AK, Gardea-Torresdey JL, Parsons JG (2007) Accumulation, speciation and cellular localization of copper in *Sesbania drummondii*. *Chemosphere* 67: 2257–2266.

Salt DE, Wagner GJ (1993) Cadmium transport across tonoplast of vesicles from oat roots. *Journal of Biological Chemistry* 268: 12297-12302.

Salt DE, Prince RC, Pickering IJ, Raskin I (1995) Mechanisms of cadmium mobility and accumulation in Indian mustard. *Plant Physiol* 109: 1427–1433.

Salt DE, Prince RC, Baker AJM, Raskin I, Pickering IJ (1999) Zinc ligands in the metal hyperaccumulator *Thlaspi caerulescens* as determined using X-ray absorption spectroscopy. *Environmental Science and Technology* 33: 712–717.

Salt DE, Blaylock M, Kumar NPBA, Dushenkov V, Ensley BD, Chet I, Raskin I (1995) Phytoremediation: a novel strategy for the removal of toxic metals from the environment using plants. *Biotechnology 13*: 468–474.

Sayers Z, Brouillon P, Vorgias CE, Nolting HF, Hermes C, Koch MH (1993) Cloning and expression of *Saccharomyces cerevisiae* coppermetallothionein gene in *Escherichia coli* and characterization of the recombinant protein. *Eur J Biochem* 212: 521–528.

Sazinsky MH, Agarwal S, Argüello JM, Rosenzweig AC (2006) Structure of the Actuator Domain from the *Archaeoglobus fulgidus* Cu^+-ATPase. *Biochemistry* 45:9949-9955.

Schmidke I, Stephan UW (1995) Transport of metal micronutrients in the phloem of castor bean (*Ricinus communis*) seedlings. *Physiol Plant* 95: 147–153.

Schat H, Llugany M, Vooijs R, Hartley-Whitaker J, Bleeker PM (2002) The role of phytochelatins in constitutive and adaptive heavy metal tolerances in hyperaccumulator and nonhyperaccumulator metallophytes. *Journal of Experimental Botany* 53: 2381–2392.

Schünemann V, Meier C, Meyer-Klaucke W, Winkler H, Trautwein AX, Knappskog PM, Toska K, Haavik J (1999) Iron coordination geometry in full-length, truncated, and dehydrated forms of human tyrosine hydroxylase studied by Mössbauer and x-ray absorption spectroscopy. *J Biol Inorg Chem* 4: 223–231.

R. Serrano (1978) Characterization of the plasma membrane ATPase of *Saccharomyces cerevisiae*. *Mol. Cell. Biochem.* 22: 51–63.

Shen ZG, Zhao FJ, McGrath SP 1997. Uptake and transport of zinc in the hyperaccumulator *Thlaspi caerulescens* and the non-hyperaccumulator *Thlaspi ochroleucum*. *Plant, Cell & Environment* 20: 898–906.

Shi J, Wu B, Yuan X, Cao YY, Chen X, Chen Y, Hu T (2008) An x-ray absorption spectroscopy investigation of speciation and biotransformation of copper in *Elsholtzia splendens*. *Plant Soil* 302: 163–174.

Siebke K, Weis E (1995a) Assimilation images of leaves of *Glechoma hederacea*: analysis of non-synchronous stomata related oscillations. *Planta* 196: 155–165.

Siebke K, Weis E (1995b) Imaging of chlorophyll-a-fluorescence in leaves: topography of photosynthetic oscillations in leaves of *Glechoma hederacea*. *Photosynthesis Research* 45: 225–237.

Smith PK, Krohn RI, Hermanson GT, Mallia AK, Gartner FH, Provenzano MD, Fujimoto EK, Goeke NM, Olson BJ, Klenk DC (1985) Measurement of protein using bicinchoninic acid. *Anal. Biochem.* 150: 76–85.

Solioz M, Vulpe CD (1996) CPx type ATPases: a class of P-type ATPases that pump heavy metals. *Trends Biochem. Sci.* 21: 237–241.

Stephan UW, Schmidke I, Stephan VW, Scholz G (1996) The nicotianamine molecule is made-to-measure for complexation of metal micronutrients in plants. *Biometals* 9: 84–90.

Stephan UW, Scholz G (1993) Nicotianamine: mediator of transport of iron and heavy metals in the phloem? *Physiol Plant* 88: 522–529.

Tomic S, Searle BG, Wander A, Harrison NM, Dent AJ, Mosselmans JFW, Inglesfield JE (2005) New Tools for the Analysis of EXAFS: The DL EXCURV Package. CCLRC Technical Report DL-TR-2005-001. *Council for the Central Laboratory of the Research Councils, Swindon, UK*

Ueno D, Ma JF, Iwashita T, Zhao FJ, McGrath SP (2005) Identification of the form of Cd in the leaves of a superior Cd-accumulating ecotype of *Thlaspi caerulescens* using 113Cd-NMR. *Planta* 221: 928–936.

Van Geen A, Adkins JF, Boyle EA, Nelson CH, Palanques A (1997) A 120 yr record of widespread contamination from mining of the Iberian pyrite belt. *Geology* 25: 291–294.

van Hoof NALM, Koevoets PLM, Hakvoort HWJ, Ten Bookum WM, Schat H, Verkleij JAC, Ernst WHO (2001) Enhanced ATP-dependent copper efflux across the root cell plasma membrane in copper-tolerant *Silene vulgaris*. *Physiologia Plantarum* 113: 225–232.

van de Mortel JE, Almar Villanueva L, Schat H, Kwekkeboom J, Coughlan S, Moerland PD, Ver Loren van Themaat E, Koornneef M, Aarts MG (2006) Large expression differences in genes for iron and zinc homeostasis, stress response, and lignin biosynthesis distinguish roots of

Arabidopsis thaliana and the related metal hyperaccumulator *Thlaspi caerulescens*. *Plant Physiol* 142: 1127–1147.

van de Mortel JE, Schat H, Moerland PD, Ver Loren van Themaat E, van der Ent S, Blankestijn H, Ghandilyan A, Tsiatsiani S, Aarts MG (2008) Expression differences for genes involved in lignin, glutathione and sulphate metabolism in response to cadmium in *Arabidopsis thaliana* and the related Zn/Cd-hyperaccumulator *Thlaspi caerulescens*. *Plant Cell Environ* 31: 301–324.

Van der Zaal BJ, Neuteboom LW, Pinas JE, Chardonnens AN, Schat H, Verkleij JA, Hooykaas PJ (1999) Overexpression of a novel Arabidopsis gene related to putative zinc-transporter genes from animals can lead to enhanced zinc resistance and accumulation. *Plant Physiology* 119: 1047-1055.

Verbruggen N, Hermans C, Schat H (2009) Molecular mechanisms of metal hyperaccumulation in plants. *New Phytologist* 181: 759-776.

Verret F, Gravot A, Auroy P, Preveral S, Forestier C, Vavasseur A, Richaud P (2005) Heavy metal transport by AtHMA4 involves the N-terminal degenerated metal binding domain and the C-terminal His11 stretch. *FEBS Lett*. 579: 1515–1522.

Verret F, Gravot A, Auroy P, Leonhardt N, David P, Nussaume L, Vavasseur A, Richaud P (2004) Overexpression of AtHMA4 enhances root-to-shoot translocation of zinc and cadmium and plant metal tolerance. *FEBS Letters* 576: 306-312.

von Wiren N, Klair S, Bansal S, Briat JF, Khodr H, Shioiri T, Leigh RA, Hider RC (1999) Nicotianamine chelates both FeIII and FeII: implications for metal transport in plants. *Plant Physiol* 119: 1107–1114.

Walker DJ, Bernal MP (2004) The effects of copper and lead on growth and zinc accumulation of *Thlaspi caerulescens* J. and C. Presl: implications for phytoremediation of contaminated soils. *Water Air Soil Pollut* 151: 361–372.

Wang J, Zhao FJ, Meharg AA, Raab A, Feldmann J, McGrath SP (2002) Mechanisms of arsenic hyperaccumulation in *Pteris vittata*. Uptake kinetics, interactions with phosphate, and arsenic speciation. *Plant Physiology* 130: 1552–1561.

Webb SM, Gaillard JF, Ma LQ, Tu C (2003) XAS speciation of arsenic in a hyper-accumulating fern. *Environmental Science and Technology* 37: 754–760.

Weber M, Harada E, Vess C, Von Roepenack-Lahaye E, Clemens S (2004) Comparative microarray analysis of *Arabidopsis thaliana* and *Arabidopsis halleri* roots identifies nicotianamine synthase, a ZIP transporter and other genes as potential metal hyperaccumulation factors *Plant Journal* 37: 269–281.

Williams LE, Mills RF (2005) P1B ATPases—an ancient family of transition metal pumps with diverse fucntions in plants. *Trends Plant Sci.* 10: 491–502.

Willmer CM, Grammatikopoulos G, Lascève G, Vavasseur A (1995) Charakterization of the vacuolar-type H^+-ATPase from guard cell protoplasts of *Commelina*. *J. Exp. Bot.* 46, 285: 383-389.

Wunderli-Ye H, Solioz M (2001) Purification and functional analysis of the copper ATPase CopA of *Enterococcus hirae*. *Biochem. Biophys. Res. Commun.* 280: 713–719.

Zhao FJ, Jiang RF, Dunham SJ, McGrath SP (2006) Cadmium uptake, translocation and tolerance in the hyperaccumulator *Arabidopsis halleri*. *New Phytologist* 172: 646-654.

Zhao F, McGrath SP, Crosland AR (1994) Comparison of three wet digestion methods for the determination of plant sulphur by inductively coupled plasma atomic emission spectrometry (ICP-AES). *Commun Soil Sci Plant Anal* 25: 407–418.

Zhou W, Hesterberg D, Hansen PD, Hutchison KJ, Sayers DE (1999) Stability of copper sulfide in a contaminated soil. *J Synchrotron Radiat* 6: 630-632.

Zimmermann M, Clarke O, Gulbis JM, Keizer DW, Jarvis RS, Cobbett CS, Hinds MG, Xiao Z, Wedd AG (2009) Metal Binding Affinities of *Arabidopsis* Zinc and Copper Transporters: Selectivities Match the Relative, but Not the Absolute, Affinities of their Amino-Terminal Domains. *Biochemistry* 48: 11640-11654.

5. Appendix

5.1. Protocol for isolation and purification of native TcHMA4 from *Thlaspi caerulescens* roots

Isolation day

1. Isolation

1.1. Before the isolation starts, the mill's detachable parts are assembled and immersed in liquid nitrogen (all parts have to be immersed to avoid tensions in the material).
A large mortar and pestle are immersed in liquid nitrogen as well and kept there for at least 10min.

1.2. Isolation buffer (330 mM mannitol (Sigma Ultra grade), 30 mM HEPES (Calbiochem Ultrol grade), 3 mM $MgCl_2$ (Merck Suprapur grade)) is prepared and cooled down to 4°C. Afterwards, 10 mM TCEP is added and dissolved by very slow stirring to minimise oxidation of TCEP. Then the pH is adjusted with KOH to 6.0 on ice. After that, "complete –EDTA" (Roche) protease inhibitor is added (one tablet of 60 mg for 50 ml of buffer).

1.3. 30 g of frozen roots are pre-crushed to pieces not larger than peas in the LN_2-cooled mortar and then ground to fine powder using the mill. This process is done twice to get powder particles as fine as possible. The mill is washed with liquid nitrogen until no more root powder comes out and the powder in the collection pot has turned into a slurry. Then 150 g of buffer are first frozen in the mortar (or, better, by pumping them as drops into a dewar with liquid nitrogen) and then ground using the mill. Again, the mill is washed with liquid nitrogen until not more buffer powder comes out and the powder in the collection pot has turned into a slurry. The buffer powder is collected in the same pot as the root powder and both should be mixed carefully.

1.4. After everything is ground, the mixture, still containing liquid nitrogen, is placed in a strongly sealed box which should be closed properly to avoid oxidation. There the material is allowed to thaw at room temperature.

1.5. When thawed completely, the mixture is transferred into ultracentrifugation tubes that were pre-cooled on ice, and centrifuged using the rotor "Type Ti 45" at 38000 rpm (113,000g) for 90 min at 4°C.

1.6. The supernatant is either taken for gels/blots or discarded. The pellet must be resuspended with another 150 ml of the above mentioned isolation buffer containing TCEP and complete. Then, the root/buffer mixture is centrifuged in the same way for 60 min again as mentioned above for washing out soluble proteins more thoroughly. The supernatant can be discarded.

2. Solubilisation

2.1. Solubilisation buffer is prepared (160 mM NaH_2PO_4 (Merck suprapur grade), 1.6 M NaCl (suprapur)) and cooled down to approx. 10°C. Then 10 mM TCEP is added and dissolved by slow stirring. Afterwards, the pH is adjusted to 6.0 and only after that, 10 mM DDM and "complete" are added. It is important to make sure that the DDM is dissolved properly, it tends to stick to the walls or lid of the glass/falcon.

2.2. The obtained root membrane pellets (1.6.) are carefully resuspended with 60 ml of solubilisation buffer using a plastic spatula. The mixture is poured into a glass bottle (100 ml) and stirred in the cold room or refrigerator (4°C) for 4-6 hours. Stirring speed should be strong enough to move the stirring bar properly, but foaming in the mixture should be avoided.

2.3. After solubilisation, the mixture is transferred into pre-cooled ultracentrifugation tubes and centrifuged with the above mentioned specifications for at least 90 min. The supernatant ("crude extract") is kept and the pellet (cell walls, etc.) discarded. A sample of the crude extract (crude1) is taken and stored in the refrigerator, not on ice as the phosphate might precipitate when kept on ice.

3. Dialysis

3.1. A piece of dialysis tube (length according to the amount of crude) is washed by putting it into double distilled water for at least 1 h, the water is exchanged from time to time.

3.2. Dialysis buffer (0.5 mM KH_2PO_4, 49.5 mM Hepes and 300 mM KCl and 100 g Chelex) is prepared and cooled down to 4°C. Then 2 mM TCEP is added, the pH is adjusted with KOH/HCl to 6.0 before 0.2 mM DDM and one tablet of complete are added and dissolved.

3.3. Then the crude extract is filled into the tube using a funnel and dialysis clamps. The filled tube is placed in 1 litre of pre-cooled (4°C) dialysis buffer. It is important to stir the buffer well at all times because Chelex that has settled down is almost impossible to remove from the bottom.

3.4. A silicon lid with attached tubes is placed on the cylinder/bottle, sealed with parafilm and used to aerate the solution with nitrogen for at least 2 min.
Then the cylinder is put on the stirring plate in the cold room over night.

4. Preparations for the following column day

4.1. Preparation of 5 l column buffer: 0.5 mM NaH_2PO_4 (Merck suprapur grade), 49.5 mM Hepes and 300 mM NaCl (suprapur). 700 ml of this buffer are needed with 0.25 M imidazole (will be adjusted to pH 9.0 on the column day, "buffer B") as well as 550 ml of this buffer (will be adjusted to pH 6.0 on the column day, "buffer C"). Store the buffers in the refrigerator or cold room (4°C).

4.2. The column has to be stored in 100 mM Ni in the night before the column run (better much longer), so it has to be nickel-loaded at least the evening before!

4.3. It has to be made sure that enough falcons and 100 ml glass bottles are available for collection of elution fractions, these should have been cleaned in 10% HCl and then stored in double distilled water for several days.

Column+characterisation day (36h)

Preparations

5.1. As TcHMA4 needs a loading and eluting temperature of 1.8°C, the column and the thermostat should be switched on first. The nickel is washed out using double distilled water. Detector and penwriter should be switched on to see what is being washed down from the column and to make sure that the recording works properly.

5.2. Buffers: The above mentioned buffers "B" and "C" both get 2 mM TCEP before the final pH is adjusted using NaOH/HCl (both suprapur): "B" to pH 9.0 and "C" to pH 6.0, on ice. Then 0.2 mM DDM is added to both of them. 2.5 l of the remaining 3.75 l column buffer get 2 mM TCEP as well, afterwards pH is adjusted with KOH to 9.0 (2-4 °C) and 0.2 mM DDM is added. The remainder of the column buffer is adjusted to pH 6 (4°C) with NaOH/HCl (both suprapur). Once the buffers are prepared, they are put on ice.

5.3. The dialysis tube is taken out of the cylinder and briefly (few seconds) immersed in a cylinder filled with double distilled water (4°C) to wash away the Chelex, then it is placed in an empty beaker and opened. A sample is taken (crude2) and stored in the fridge.

5.4. Then, the crude extract is diluted to 250 ml with the remainder of the 5 l column buffer (mentioned above pH 6.0) on ice. Centrifuge again for 50 min (specifications as mentioned above) to avoid precipitation of protein on the column pre-filter.

5.5. Once the crude is centrifuging, the column has to be closed (detached from the tubes) and the column tubes are washed with the according buffer. This should be started with tube C and finished with tube A. Then the column is connected again and equilibrated for at least 10 min (max. 30min) with buffer A at a flow rate of 3 ml.min^{-1}.

5.6. The crude 3 extract is adjusted to pH 9.0 at 2-4 °C in a beaker. Only after pH adjustment, the solution is transferred into a 250ml glass cylinder which is then placed in ice. After that, a sample (crude 3) is taken and stored in the fridge.

5.7. The protein concentrators need to be washed with 10 ml of double distilled water before they are ready to use, to do so, they should run in the table centrifuge for 5min at 3000g.

Column run

6.1. As soon as the baseline of the column is stable, the protein is loaded over a special tube (connected to tube A over a valve) with a flow rate of 3 ml/min and the "flow through" is collected, a part of it will be concentrated later(so called W_0).

6.2. Once all protein is loaded onto the column, the post-wash is started with buffer A at a flow rate of 3 ml/min. As soon as the loading peak has dropped to half of its maximal height, the flow rate is increased to 9 ml/min. The post-wash is continued until a stable baseline is reached (typically 2 h).

6.3. Then the gradient (currently saved as Nr. 2 in the controller unit) is started, the beginning is marked on the writer sheet. The gradient takes 168 min and first mixes buffers A and B to increase the content of imidazole at a constant pH of 9.0, then buffers B and C are mixed for a constant concentration of imidazole and decreasing pH.
Peaks are collected manually.

6.4. When the gradient has ended, elution is continued with 100% buffer C at 9 ml/min (=final condition of the gradient) until the baseline is reached. This part of the last peak is often the most pure fraction of TcHMA4!

6.5. Afterwards, the pump is supplied with water instead of buffer, the thermostat is switched off, and regeneration of the column is started.

6.6. The collected fractions ("elutions") are adjusted to pH 6.0 immediately after coming out of the column, two samples of 500 μl are taken and stored on ice (for precipitation gel and blot, see end of protocol), and the volume of the elutions is measured. They are divided for EXAFS (2/3 of the volume, this part is concentrated with cadmium or zinc) and activity test (1/3 of the volume, strictly to be kept metal free!).

6.7. Concentration of W_0 and the elutions takes place at 4°C and 3000 g. Every time when new sample is added to the concentrator, the protein close to the membrane should be pipetted up and down carefully, avoiding aeration and not touching the membrane with the pipet tip as the membrane might be damaged.

The final volume of the concentrated fractions is calculated as follows:

0.018 x fraction volume

6.7. The weight of the peaks needs to be estimated; therefore the writer sheet is scanned and set to 200% and printed out on 80 g.m^{-2} paper. The baseline has to be drawn through the whole gradient. The peaks are now cut out and weighted for each fraction.

6.8. Samples for activity test are taken now. The pipetting plan includes volumes as follows:

$$\mu l\_volume\_for\_activity\_test = \frac{0.286 * ml\_volume\_before\_concentration}{mg\_peak\_weight}$$

Activity assay

Pipeting plan:

A) Reaction samples

Phosphate Standards [µM]	Phosphate [µl] (stock: 4000 µM)	ddH$_2$O [µl]
0	0	100
100	2.5	97.5
200	5	95
300	7.5	92.5
400	10	90
500	12.5	87.5
600	15	85
800	20	80
1000	25	75
1200	30	70
1600	40	60
2000	50	50

Protein samples:

0 (blank)	E1 0.1 µM Zn Stock 2 µM	E1 0.3 µM Zn Stock 6 µM	E1 1 µM Zn Stock 20 µM	E1 3 µM Zn Stock 60 µM	E1 10 µM Zn Stock 200 µM
	E1 0.1 µM Cd Stock 2 µM	E1 0.3 µM Cd Stock 6 µM	E1 1 µM Cd Stock 20 µM	E1 3 µM Cd Stock 60 µM	E1 10 µM Cd Stock 200 µM
	E1 0.1 µM Cu Stock 2 µM	E1 0.3 µM Cu Stock 6 µM	E1 1 µM Cu Stock 20 µM	E1 3 µM Cu Stock 60 µM	E1 10 µM Cu Stock 200 µM

E= Elution

Pipet in all blue samples 70 µl ddH$_2$O and 20 µl ATPase test-solution (sample E1 + "0": 170 µl ddH$_2$O instead of 70 µl)

Pipet in all protein samples that include metal (red samples) 10µl of the according metal stock solution, 160 µl ddH$_2$O and 20 µl ATPase test-solution

The 0 phosphate standard is repeated each time before starting with the measuring of a new protein sample. It is pipetted like the „normal" 0 phosphate standard.

At the end, there has to be 190 µl liquid in each reaction vial.

B) Reconstitution samples
The incubation oven needs to be set to 30°C.

B1) Make one reconstitution batch for the standards and one each for the column fractions. Take the following volumes, which are calculated per reaction vial, times the number of vials + 20%.

B2) Pipet per sample 9 µl minus the calculated volume of the protein fraction of LEW-buffer pH 6 (= column buffer A after adjustment to pH 6).

B3) Add 1 µl of lipid per sample.

B4) Pipet per reaction vial the protein volume calculated for the activity test, and gently mix the reconstitution batches by inverting them a few times.

B5) As soon as the protein in added, the samples are incubated 20 min at 30°C in the oven.

B6) After incubation, pipet WITH LOW-RETENTION TIPS 10 µl of the matching reconstitution batches into the vials with the reaction mixes. Then 20 µl ATP (50 mM Mg-ATP or Na_2-ATP) is added with low retention tips. The samples are incubated 40 min at 30°C.

B7) After incubation, the ATPase reaction is stopped with 800 µl stop solution. After stopping the reaction, 40 µl 10%ig ascorbic acid is added to all samples for colour development. After exactly (!) 2 min, 35 µl 34% 3Na-citrate is added to stop the reaction.

B8) The OD is measured at a wavelength of 750nm with UV/VIS spectroscopy. During the calibration series and while changing from one to another sample of the same protein fraction, only empty the cuvette, do not wash it, because the error caused by residual contents is in this case less than the error that would be caused by residual water.

Buffers needed (all made from Merck Suprapur or equivalent chemicals!):

Lipid stock solution: 6.6 mg lipid is dissolve in 10 ml ddH$_2$O (has to be prepared freshly each day before use!)

Vitamin C stock solution: 2 g Vit C in water, filled to 20 ml (has to be prepared freshly each day before use!)

ATPase test solution: 600 mM TRIS, 2 M NaCl, 100 mM MgCl$_2$, pH 8.0 (can be stored in the refrigerator for several weeks)

Stop solution: 0.5% SDS, 0.5% NH$_4$Mo(H$_2$O)$_4$, 2% sulfuric acid (can be stored many months at 4°C)

LEW buffer: buffer A from column run, adjusted to pH 6 with HCl (has to be prepared fresh after each column run, so that the TCEP in this buffer has the same oxidation state as in the protein samples).

Na$_3$-Citrate: 34% (can be stored many months at 4°C)

Modifications for temperature-dependent activity test

1) Each temperature has its own row of standards (i.e. typically four standards plus 6 samples for a 10-well row of the gradient thermostat)
2) After pipetting the vials with the reaction mix (at room temperature), they are pre-incubated at the reaction temperature for at least 10min before adding the protein. This means placing them in the thermostat either directly after the reaction mix was completed, or (if there is a longer gap before work is continued so that the reaction mix is stored in the fridge) immediately after the reconstitution mix has been placed in the 20 min 30°C incubation step for reconstitution.
3) The same amount of concentrated (by the same factor as the protein samples to have the same concentration of detergent) LEW buffer pH6 is needed in the reconstitution mix of the standards as the concentrated protein fractions in the reconstitution mix are used for the samples.
4) Always row at highest temperature is pipetted first and then the lower temperatures. In this way, the faster passive ATP degradation at higher temperatures (also after the "stop" solution is added!) is compensated by the longer time it takes until the low temperature samples are developed with vitamin C + citrate.

UV/VIS and EXAFS spectroscopy

The samples for spectroscopy (with and without cadmium) are taken and 70 µl cuvettes are used. The scan runs over a wavelength from 250 to 800 nm with 0.2 nm sampling interval and 0.5 nm spectral bandwidth, reference is taken with buffer concentrated without protein.

The samples with cadmium are further concentrated in microcon concentrators (exclusion size 30 kD) to a final volume of ca. 35 µl, which is then transferred with a syringe (the needle needs to fit into the cuvette hole, this has to be checked before the protein is transferred into the syringe!) into labelled EXAFS cuvettes (the manual by Wolfram Meyer-Klaucke gives more information on how to fill these cuvettes successfully) and shock-frozen in supercooled (-140°C) isopentane.

Precipitation

1) TCA 20% and acetone need to be p.a. precooled on ice
2) The doubled volume of TCA is added to the protein and mixed immediately by vortexing.
3) The protein-TCA mixture is centrifuged at least 60 min with 16,000g at 1°C (table centrifuge)
4) The supernatant must be taken out completely (important: it helps to suck out again after about 1 min with a 200 µl pipet tip to remove all residues of acid)
5) Washing step with the same volume of acetone
5) The samples are centrifuged for 5 min, again the supernatant must be taken out completely
6) The pellet is then dissolved in 33% sample buffer/ 66% water mixture
7) If the sample turns yellowish, a very small amount (start with 0.1µl) of low concentrated NaOH should be added to adjust to neutral pH (--> solution turns blue)

Gel recipe for 8% SDS gels

For two small (10x10 cm, 0.8 mm thick) gels

Resolving gel preparation
Acrylamide: 2.87 ml
Resolving gel buffer: 2.5 ml
ddH$_2$0: 4.43 ml
APS (10% in dd water): 70 µl

Stacking gel preparation
Acrylamide: 650 µl
Stacking gel buffer: 4.35 ml
APS (10% in dd water): 50 µl

Resolving gel buffer:
1.5 M Tris-HCl, 0.4% (v/v) TEMED and 0.4% (w/v) SDS; pH9.0
Stacking gel buffer:
0.1M Tris-HCl, 0.11% (v/v) TEMED and 0.11% (w/v) SDS; pH6.8

Blotting

Transfer buffer for western blot:
3.2g TRIS, 17.3 g Glycine, 300 ml methanol for 1.2L buffer

The buffer should be precooled by putting the blotter with the buffer on ice at least 3 h before the run.

Nitrocellulose membrane and filterpapers (2 large papers from the box cut into 4 pieces each for two blotting cassettes) are prewet in the cold transfer buffer.

The blotting assembly should be done as follows:
bright side of the cassette - sponge - two filter papers - nitrocellulose - gel (there must not be any bubbles between gel and membrane) - two filter papers – sponge - dark side of the cassette

The cassette needs to be closed properly but without too much force, as the clasps tend to break easily.

Blotting time 1 h at 200 mA with the membrane on the bright side of the cassette, this bright side must be placed towards the red cable.

After blotting, the blotter is switched off and the cassettes are openend. The membranes should be washed with PBST for 5 min before putting them into blocking solution (10% Roti-Block in PBST) for at least 1 hour.

2-3x short washes with PBST

Incubation with primary antibody at room temperature (20 µl primary in 18 ml PBST with 2 ml Roti-Block) for at least 6 h, better 10 h.

Washing steps 2-3x with PBST for 5 min.

Incubation with secondary antibody for 1 h (18 ml PBST and 2 ml Roti-Block with 17 µl of a 1:10 dilution of secondary AB)

Washing steps 2-3x for 5 min in PBST and once in AP-Buffer for 5 min.

Development of the blot in 20 ml of AP-Buffer with one aliquot of NBT and one aliquot of BCIP (-20°C). Once clear bands are visible, the development can be stopped by washing the blot with tap water several times. Incubation in tap water will slowly increase the signal.

Scanning and drying of the blots.

Silver Staining

1. Fixation of the gel with fixing solution (50 ml Methanol, 10 ml acetic acid, 40 ml dd water and 50 µl formaldehyde) over night.

2. Distilled water wash 2x

3. Incubation for 1 h in 50% Ethanol for dehydration – the gel shrinks.

4. Sodium thiosulphate treatment for 45 s (40 mg in 100 ml dd water)

5. dd water wash at least 3x, better more often

6. Silver nitrate incubation for 20 min (200 mg in 100 ml dd water plus 75 µl formaldehyde), covered in aluminium foil on the shaker

7. Collection of the silver solution (metal waste) - dd water wash at least 3x, better more often

8. Development with 4.5 g sodium carbonate dissolved in 100 ml of dd water and 50 µl formaldehyde. The bands can appear very quickly, the gel should not stay unattended during the staining as the gel turns black very quickly. As soon as clear bands are visible, the reaction is stopped by pouring out the solution and replacement with stop solution (50 ml methanol, 10 ml acetic acid, 40 ml dd water)

5.3. Author contributions

This thesis was written by Barbara Leitenmaier using only the sources mentioned in the text and listed in the reference section (4.). The author contributed to the specific chapters as follows:

Chapter 2.1.
Experiments were planned by HK, for the first series of experiments together with IŠ. Plants for the first series of experiments were cultivated and measured by HK. The FKM has been developed by HK and MT. In the later series of experiments, plant care and microscopic measurements have been carried out by AP and BL, the quantification of chlorophyll in plant extracts has been done by HK and AP. Qualitative visualization using a fluorescent dye was done by HK and BL. The later series of experiments have been analysed by HK, AP and BL, statistics and graphs were done by HK as well as writing of the manuscript. All authors revised the manuscript.

Chapter 2.2.
Experiments were planned by HK and BL. Plant cultivation and measurements were done by BL. Analysis of data has been carried out by BL and HK, as well as writing of the manuscript and statistical analysis.

Chapter 2.3.
Experiments were planned by HK; AP and BL. Plant cultivation was done by AP, BL and MY. Experiments have been carried out by HK, AP, BL and MY with the help of GL. LK provided the specific antibody for TcHMA4 and WW and PK contributed with fruitful discussions. The manuscript has been written by AP, HK and BL and has been revised by all authors.

Chapter 2.4./2.5.
Experiments were planned by HK and BL. Plant cultivation has been done by BL. Isolation and purification of TcHMA4 were carried out by BL and AWitt. Activity assays were prepared and measured by HK, BL, AS and AWitzke. EXAFS samples were prepared by HK and BL, WMK measured the first EXAFS sample in 2007, HK and BL measured the second sample in 2010. Analysis of EXAFS data has been performed by WMK and HK and BL.
PK contributed with fruitful discussions. The manuscript has been written by BL and was revised by HK.

Chapter 2.6.

Experiments have been planned by HK. Plant cultivation was done by BL and BG, as well as measurements of chlorophyll fluorescence kinetics. Analysis of the FKM data was done by HK, BL and BG. Quantification of chlorophyll was done by HK. Preparation of plant samples for EXAFS measurements was done by HK. EXAFS measurements were performed by HK together with WMK; AM, HK and WMK did the analysis of the EXAFS data. Copper complexes for spectroscopic measurements were prepared by BL, BG and HK. AAS measurements were performed by HK. EPR measurements and data analysis were done by PK. The manuscript was written by AM and HK and revised by all authors.

5.4. Acknowledgements

First of all, I thank my supervisor Hendrik Küpper for his continuous support during all the years. He was always there to listen, to discuss and to give advices. He showed me different ways to approach research questions and also, the need to be persistent if things do not work out exactly as planned. He gave me the opportunity to attend many conferences, see different countries and meet important people, and, most importantly, with his enthusiasm for his field he is responsible for involving me in the topic "metals and plants" and I am deeply grateful for that!

I also thank Peter Kroneck for supporting and encouraging me. He gave me very good ideas and during our lunch and coffee breaks I learned more than in any lecture. Special thanks goes to him for giving me the opportunity to travel to Mexico for a summer school in 2008, were I not only learned a lot about metals in biology, but met great people in a beautiful country.

Thanks also go to Wolfram Welte for financial support during the first months of my PhD project and for interesting discussions about ATPases in general, detergents in particular, and many other important topics.
Wolfram Meyer-Klaucke was involved in measurements of EXAFS, in 2007 at the Desy in Hamburg where he arranged beam time and helped with data recording and analysis and in 2010 at the SSRL in Stanford, CA, USA, where he managed the application for beam time and introduced me to several important people.

As successful labwork is not possible without a good atmosphere in the lab, thanks to all the current and former members of the Küpper group, in particular Aravind Parameswaran, who started the work on TcHMA4 with me; George Thomas, Elisa Andresen and Seema Mishra, sharing the lab with you was great!
Thanks also to the students who were working on TcHMA4 with me: Anastasia Stemke, Annabell Witzke, Birte Baudis, Nils Schoelzel, Michael Maier and Timo Witt.
The gardeners Anne Kern and Otmar Ficht helped with harvesting of seeds and good advice concerning plant care, thank you!
Special thanks go to the best Hiwi ever, Annelie Witt. She spent many nights in the lab with me and without her, the work on TcHMA4 would have been much more difficult and also very importantly, less fun!

I am deeply grateful to my mother Maria for supporting and encouraging me through all my life and also for believing in me, when I myself wasn´t able to do so.

My aunt Claudia was always interested in my work and supported me financially, thank you so much for that!

And last but not least, thanks to Andreas, my best friend and partner. He was always there for me, supported my ideas (and still does!) and helped me through difficult days. Thank you so much for everything!

i want morebooks!

Buy your books fast and straightforward online - at one of world's fastest growing online book stores! Environmentally sound due to Print-on-Demand technologies.

Buy your books online at
www.get-morebooks.com

Kaufen Sie Ihre Bücher schnell und unkompliziert online – auf einer der am schnellsten wachsenden Buchhandelsplattformen weltweit! Dank Print-On-Demand umwelt- und ressourcenschonend produziert.

Bücher schneller online kaufen
www.morebooks.de

VDM Verlagsservicegesellschaft mbH
Heinrich-Böcking-Str. 6-8   Telefon: +49 681 3720 174   info@vdm-vsg.de
D - 66121 Saarbrücken   Telefax: +49 681 3720 1749 www.vdm-vsg.de

Printed by Books on Demand GmbH, Norderstedt / Germany